D0212422

STEVE JOBS

Recent Titles in Greenwood Biographies

Ulysses S. Grant: A Biography
Robert B. Broadwater

Thurgood Marshall: A Biography
Glenn L. Starks and F. Erik Brooks

Katy Perry: A Biography
Kimberly Dillon Summers

Sonia Sotomayor: A Biography
Meg Greene

Stephen Colbert: A Biography
Catherine M. Andronik

William T. Sherman: A Biography
Robert P. Broadwater

Pope Benedict XVI: A Biography
Joann F. Price

Pelé: A Biography
Lew H. Freedman

Malcolm X: A Biography
A.B. Assensoh and Yvette M. Alex-Assensoh

Elena Kagan: A Biography
Meg Greene

Bob Dylan: A Biography
Bob Batchelor

Jon Stewart: A Biography
Michael Blitz

Bill Gates: A Biography
Michael B. Becraft

Clint Eastwood: A Biography
Sara Anson Vaux

STEVE JOBS

A Biography

Michael B. Becraft

GREENWOOD BIOGRAPHIES

An Imprint of ABC-CLIO, LLC
Santa Barbara, California • Denver, Colorado

Copyright © 2017 by ABC-CLIO, LLC

All rights reserved. No part of this publication may be reproduced, stored in a retrieval system, or transmitted, in any form or by any means, electronic, mechanical, photocopying, recording, or otherwise, except for the inclusion of brief quotations in a review, without prior permission in writing from the publisher.

Library of Congress Cataloging-in-Publication Data

Names: Becraft, Michael B., author.
Title: Steve Jobs : a biography / Michael B. Becraft.
Description: Santa Barbara, California : Greenwood, [2017] | Series: Greenwood biographies
Identifiers: LCCN 2016031232 (print) | LCCN 2016046038 (ebook) | ISBN 9781610694964 (hard cover : alk. paper) | ISBN 9781610694971 (ebook)
Subjects: LCSH: Jobs, Steve, 1955-2011. | Apple Computer, Inc. | Computer engineers—United States—Biography. | Businesspeople—United States—Biography. | Computer industry—United States—History.
Classification: LCC QA76.2.J63 B43 2017 (print) | LCC QA76.2.J63 (ebook) | DDC 338.7/6100416092 [B]—dc23
LC record available at https://lccn.loc.gov/2016031232

ISBN: 978-1-61069-496-4
EISBN: 978-1-61069-497-1

21 20 19 18 17 1 2 3 4 5

This book is also available as an eBook.

Greenwood
An Imprint of ABC-CLIO, LLC

ABC-CLIO, LLC
130 Cremona Drive, P.O. Box 1911
Santa Barbara, California 93116-1911
www.abc-clio.com

This book is printed on acid-free paper ♾

Manufactured in the United States of America

CONTENTS

SERIES FOREWORD

In response to school and library needs, ABC-CLIO publishes this distinguished series of full-length biographies specifically for student use. Prepared by field experts and professionals, these engaging biographies are tailored for students who need challenging yet accessible biographies. Ideal for school assignments and student research, the length, format, and subject areas are designed to meet educators' requirements and students' interests.

ABC-CLIO offers an extensive selection of biographies spanning all curriculum-related subject areas including social studies, the sciences, literature and the arts, history and politics, and popular culture, covering public figures and famous personalities from all time periods and backgrounds, both historic and contemporary, who have made an impact on American and/or world culture. The subjects of these biographies were chosen based on comprehensive feedback from librarians and educators. Consideration was given to both curriculum relevance and inherent interest. Readers will find a wide array of subject choices from fascinating entertainers like Miley Cyrus and Lady Gaga to inspiring leaders like John F. Kennedy and Nelson Mandela, from the greatest athletes of our time like Michael Jordan and Muhammad Ali to the most amazing success stories of our day like J.K. Rowling and Oprah.

While the emphasis is on fact, not glorification, the books are meant to be fun to read. Each volume provides in-depth information about the subject's life from birth through childhood, the teen years, and adulthood. A thorough account relates family background and education, traces personal and professional influences, and explores struggles, accomplishments, and contributions. A timeline highlights the most significant life events against an historical perspective. Bibliographies supplement the reference value of each volume.

PREFACE

One of the lessons I impress upon students while teaching leadership courses is that a person's leadership qualities are not defined by his or her self-perception; they are measured by how one is perceived by others. In this book, I do take care to not only use the words of Steve Jobs but also the words of his peers, both contemporaneously and after his passing.

While Steve Jobs is undoubtedly linked to Apple Computer, NeXT, and Pixar, note that the process of writing a biography is not to provide a history of an individual's employment, other than the parts absolutely necessary to provide information to the reader about the life of Mr. Jobs. There are exceptional books about each of those firms written for the same audience. When innovations such as Apple II, Macintosh, iPod, iPhone, and iPad are addressed in this book, note that the history of these devices will be used to more fully describe Mr. Jobs and his influence on technology. So while there won't be schematics or diagrams of the technology, there will be discussion of the changes in the world that resulted from Mr. Jobs's influence.

An enigmatic leader who profoundly changed over time, he has been the subject of copious articles, interviews, and profiles. In fact, this book incorporates five hundred endnotes to help the audience appreciate his influences, life, and work.

Readers should be very clear that Jobs was a man of contrasts—removed from his company Apple for not being a sophisticated executive, he came back with a vision that clearly saved the company. Between his two stints at Apple, he founded two technology firms that still have very clear influences today. And he was exactingly focused on design as well as function, consistently speaking of the potential for the consumer who would use the products created by his firms.

Personally and professionally, Jobs had exceptionally high expectations. He was also abrasive and brusque in situations where that behavior might not have been necessary. Opportunistic, he could be harmful to a wide range of individuals, including his closest business partner.

ACKNOWLEDGMENTS

I continue to be exceptionally fortunate in the opportunities provided to me, opportunities that are not available to the majority of individuals throughout the world. Based upon the success of my previous Greenwood biography of Bill Gates, I am happy to have been invited to write a biography of Steve Jobs as well. While this work contains copious references to help high school and college students learn about Steve Jobs, I hope this work is also enjoyable for a wider audience.

Much of this work was completed at Naval Support Activity Bethesda / Walter Reed National Military Medical Center—the assistance of the personnel here on everything except writing has been greatly appreciated.

I have a lifelong fear of missing someone in acknowledgments; while I don't identify names, I am always exceptionally grateful for the way I have been shaped over the years. I even have gratitude for Mr. Jobs; although I am a hardware-agnostic individual, much of the computing I did over a 15-year span was done on Apple II, NeXT, and Macintosh products. The NeXT machines I was able to use my first few years of college—when compared to the Windows alternatives—were indeed revolutionary. And my favorite video game as an elementary school student was constructed by the pairing of Jobs and Woz.

Mr. Jobs believed the best gift he could give those he did not know were elegant devices that would provide increasing levels of utility; while he was not extensively part of many charitable activities like many other exceptionally wealthy individuals, each of us can focus on improving the lives of others in the best way we can.

TIMELINE: EVENTS IN THE LIFE OF STEVE JOBS

1955 Steven Paul Jobs is born, adopted by Paul and Clara Jobs
1957 Paul and Clara Jobs adopt Patricia (Patty) Jobs
1963 Future wife Laurene Powell is born on November 6
1968 Summer job with Hewlett Packard at age 13
1969 Steve Jobs meets Steve Wozniak
1972 Jobs and Wozniak hack pay telephones with Blue Boxes; Jobs drops out of Reed College, continues to attend classes
1974 Travels to India and begins work at Atari
1975 Jobs is arrested in Eugene, Oregon, for an outstanding speeding ticket
1976 Apple Computer Company is founded as a partnership on April 1, 1976
1977 Apple Computer, Inc. is begun as a corporation on January 3, 1977
1978 Lisa, his first daughter, is born by Chrisann Brennan on May 17, 1978
1980 Apple IPO makes Jobs worth more than $100 million
1983 Jobs lures PepsiCo CEO John Sculley to become leader of Apple

1984 Macintosh is launched after the "1984" Advertisement

1985 Jobs resigns from Apple on September 17; Starts NeXT after leaving Apple

1986 Purchases Pixar from LucasFilm; Steve Jobs meets biological sister Mona Simpson

1989 Named *Inc. Magazine*'s Entrepreneur of the Decade; Meets Laurene Powell

1990 Proposes to Laurene Powell

1991 Marries Laurene Powell on March 18, 1991; Son Reed Paul Jobs born September 1991

1993 Father Paul Jobs dies

1995 Daughter Erin Siena Jobs born in August

1996 Sister Mona Simpson's book *A Regular Guy*, patterned after her brother, is released

1997 Apple purchases NeXT, bringing Jobs back to Apple

1998 Daughter Eve is born

2003 Initially diagnosed with cancer

2004 Jobs has a cancerous tumor removed in July

2006 Pixar sells to Disney, making Jobs the largest Disney shareholder

2009 Jobs has a liver transplant

2011 Steve Jobs dies October 5, 2011

Part 1

EARLY LIFE

The story of Steve Jobs begins not in the United States, but in Syria. Although Jobs always stressed that his birth parents were Paul and Clara Jobs (and they are in fact listed on his birth certificate as his parents), Jobs's biological parents were Abdulfattah Jandali (of Homs, Syria) and Joanne Schieble (of Wisconsin). But of course to Jobs, his parents were always Paul and Clara.

Chapter 1

UPBRINGING, FAMILY, AND EDUCATION

Steve Jobs was adopted by Paul and Clara Jobs. His adoptive parents made a commitment to Jobs's biological mother that he would be required to attend college (although Paul and Clara had not). The family was middle-class but provided Jobs with everything he ever wanted in life. Paul Jobs also helped develop his son's interests in electronics. Jobs's father was a machinist for Spectra Physics, a company that built lasers; his mother was a clerk for Varian, at that time a manufacturing company. Steve Jobs would describe his dad as buying cars at a junkyard for $50, and then selling them to young adults for Jobs's "college fund," to help fulfill the promise to Jobs's biological mother.[1]

> I was very lucky. My father, Paul, was a pretty remarkable man. He never graduated from high school. He joined the Coast Guard in World War II and ferried troops around the world for General Patton; and I think he was always getting into trouble and getting busted down to Private. He was a machinist by trade and worked very hard and was kind of a genius with his hands. He had a workbench out in his garage where, when I was about five or six, he sectioned off a little piece of it and said "Steve, this is your workbench now." And he gave me some of his smaller tools and showed me

> how to use a hammer and saw and how to build things. It really was very good for me. He spent a lot of time with me.
>
> —Steve Jobs[2]

His parents always encouraged his interests. At an early age, Jobs began to express an interest in technology like his father.

> They encouraged my interests. My father was a machinist, and he was a sort of genius with his hands. He can fix anything and make it work and take any mechanical thing apart and get it back together. That was my first glimpse of it. I started to gravitate more toward electronics, and he used to get me things I could take apart and put back together.
>
> —Steve Jobs[3]

Paul Jobs—a builder of lasers—instilled in Steve the concept that all aspects of a project required the same amount of attention. As an example, he used a fence. The unseen parts of the fence required the same level of attention as the parts that were clearly visible. In future projects, Jobs would insert detail in components and parts that the average user would never see.

> You got to make the back of the fence that nobody will see just as good looking as the front of the fence. Even though nobody will see it, you will know, and that will show that you're dedicated to making something perfect.
>
> —Paul Jobs[4]

Paul Jobs was enough of a technical tinkerer to show Steve that sometimes one would undertake projects simply for the sake for completing the project. Paul Jobs even constructed a laser inside his car's engine compartment as early as the 1970s.

> He had his Ford Ranchero with a laser beam installed in the engine compartment, and he had a big toggle switch on the dashboard that said "laser." I thought that was wonderful. That's Silicon Valley for you. So his father was a tinkerer.
>
> —Dan Kottke[5]

Even Jobs's future business partner, Steve Wozniak, recognized how much Paul Jobs taught Steve about technology at a very early age.

> His father was a good technical father … He liked to show him how things worked. I know that Steve looks back and thinks that he didn't appreciate him enough.
>
> —Steve Wozniak[6]

There is no doubt that Paul was an excellent father to young Steve. In his early 40s, Jobs would say that he had just one objective when interacting with his own children, although many, including one of his own children, would state he wasn't always a perfect reflection of Paul Jobs.

> Just to try to be as good a father to them as my father was to me. I think about that every day of my life.
>
> —Steve Jobs[7]

For those who made the mistake of referring to Paul and Clara as his "adoptive parents," Jobs removed all ambiguity. Jobs held a firm belief that Paul and Clara Jobs were his true parents. A mention of his "adoptive parents" is quickly cut off. "They were my parents," he says emphatically.[8]

As Junod would later state, "he is an adoptive parent's fondest dream, not because he hit the jackpot and became a billionaire but rather because he has always regarded his adoptive parents, the late Paul and Clara Jobs, as his real parents, pure and simple."[9]

ELEMENTARY SCHOOL WAS A CHALLENGE

Jobs would describe elementary school as being a challenge for him. Part of the challenge was that his doting father Paul had allowed him to explore his interest in electronics, and his mother Clara had taught him to love reading before he went to school. School had authority figures and Jobs was not allowed to undertake the activities he desired.

> School was pretty hard for me at the beginning. My mother taught me how to read before I got to school and so when I got

> there I really just wanted to do two things. I wanted to read books because I loved reading books and I wanted to go outside and chase butterflies. You know, do the things that five-year-olds like to do. I encountered authority of a different kind than I had ever encountered before, and I did not like it.
>
> — Steve Jobs[10]

Jobs paired with a friend, Rick Farentino, to cause trouble at school. One day, the two talked to their classmates, traded bike lock combinations, and then managed to put incorrect locks on every bike at the school. Switching locks on bikes didn't destroy school property, but Jobs and Farentino were also involved with property damage.

> We set off explosives in teacher's desks. We got kicked out of school a lot.
>
> — Steve Jobs[11]

In third grade, the pair was actively placing explosives in the desks of teachers. The school decided that Jobs and Farentino could never be allowed in the same class again. Jobs cited his fourth-grade teacher as getting him excited about learning again, giving him challenges, paying him for completing math workbooks, and even buying him kits so he could undertake projects like building cameras.

> It was really quite wonderful. I think I probably learned more academically in that one year than I learned in my life. It created problems though, because when I got out of fourth grade they tested me and they decided to put me in high school and my parents said "No." Thank God. They said, "He can skip one grade but that's all." ... And I found skipping one grade to be very troublesome in many ways. That was plenty enough. It did create some problems.
>
> — Steve Jobs[12]

ADOPTED SISTER—PATRICIA JOBS

Jobs also had an adopted sister, Patricia, who joined the family in 1957. She received the same love from Paul and Clara, although the personalities of Steve and Patricia never meshed. "Steve had never gotten along that well or felt particularly close to his younger sister Patti, who was also adopted. Paul and Clara Jobs had wanted children so much that they adopted two of them, and as parents they were unfailingly loving, supportive, and self-sacrificing."[13]

Patricia would later become one of the first part-time workers at Apple. She was paid $1 for each board she assembled as part of the earliest kit computers. After the deaths of Clara and Paul Jobs, Patti would come to own, and still does own, the home where Apple was founded.

MEETING BILL FERNANDEZ

Paul's transfer to Cupertino for work was beneficial for young Steve Jobs. It was in Cupertino where he met his good friend (and future Apple employee) Bill Fernandez in middle school. Even then, Jobs's refusal to conform to traditional social norms shaped the individuals with whom he would interact best.

> We were both nerdy, socially inept, intellectual and we gravitated towards each other. We both also were not at all interested in the superficial bases upon which the other kids were basing their relationships, and we had no particular interest in living shallow lives to be accepted. So we didn't have many friends.
>
> —Bill Fernandez[14]

Fernandez also mentioned that he and Jobs spent time walking around the neighborhood and having extensive talks, a pattern Jobs would follow for the rest of his life. Many of Jobs's major friendships—and major decisions—were made or sustained through these walks.

> He and I also spent endless hours walking around the neighborhood, particularly in some of the nearby, undeveloped wild lands, talking about life, the universe, and everything.
>
> —Bill Fernandez[15]

Meeting Bill Fernandez was also a precursor to starting Apple; Steve Jobs and Steve Wozniak would initially meet through Fernandez.

Fernandez's family lived near Wozniak, and Bill was involved with the first computer that the older Steve Wozniak built (Jobs was not). Referred to as the "Cream Soda" computer, it was so named because Wozniak and Fernandez would frequently take breaks for cream sodas while building it.[16] The lifelong friendship and intermittent business relationships between Jobs and Wozniak began in 1968 after their introduction by Fernandez.

WORKING AT HEWLETT-PACKARD

In the summer between junior high and high school, Jobs got a job with Hewlett-Packard after calling Bill Hewlett to simply request parts for a project. At the time, Mr. Hewlett's number was listed in the telephone directory. Jobs was young at the time, as he had skipped fifth grade. The simple act of being involved in creating electronic components made him happy, given his experience tinkering with his father.

> When I was 12 or 13, I wanted to build something and I needed some parts, so I picked up the phone and called Bill Hewlett—he was listed in the Palo Alto phone book. He answered the phone and he was real nice. He chatted with me for, like, 20 minutes. He didn't know me at all, but he ended up giving me some parts and he got me a job that summer working at Hewlett-Packard on the line, assembling frequency counters. Assembling may be too strong. I was putting in screws. It didn't matter; I was in heaven.
>
> —Steve Jobs[17]

DISILLUSIONMENT WITH RELIGION

At the age of 13, Jobs became disillusioned with formal religion. Paul and Clara wanted him to be exposed to church, so the family frequently attended the local Lutheran church. He had seen the July 1968 *Life* magazine cover featuring an article, "Starving Children of Biafra War." Biafra was a region of southeast Nigeria, which has since been reabsorbed into Nigeria. Jobs's worry about the magazine cover and article led him to ask the pastor of the Lutheran church about the article. He would later describe that interaction to Walter Isaacson, who provided the full narrative:

> "Even though they were not fervent about their faith, Jobs's parents wanted him to have a religious upbringing, so they took him to the Lutheran church most Sundays. That came to an end when he was thirteen. In July 1968 *Life* magazine published a shocking cover showing a pair of starving children in Biafra. Jobs took it to Sunday school and confronted the church's pastor. "If I raise my finger, will God know which one I'm going to raise even before I do it?"
>
> The pastor answered, "Yes, God knows everything."
>
> Jobs then pulled out the *Life* cover and asked, "Well, does God know about this and what's going to happen to those children?"
>
> "Steve, I know you don't understand, but yes, God knows about that."
>
> Jobs announced that he didn't want to have anything to do with worshipping such a God, and he never went back to church."[18]

After high school, Jobs would make a pilgrimage to India and adopt Zen Buddhist principles he would use to shape his belief system over the remainder of his life. The Buddhist principles did not require him to believe in the God he had rejected as an early teenager. However, his future behaviors may not have always been consistent with those Buddhist beliefs he elected to adopt.

BLUE BOXES

Blue boxes worked by precisely creating all the tones used by the phone company to connect telephone calls. In order to understand why a blue box might be desirable to own, one has to think about the prevalence of telephones, and types of telephones, available in 1971. Many homes lacked phones and cell phones had not been invented. This meant that when an individual was at any location other than home or work, the only option for telephone calls would be a pay telephone.

Pay telephones had small costs for local calls but long-distance calls from a pay phone could cost many dollars and were charged per minute.

For a college student or someone travelling, the cost of a blue box might be recouped in as little as a few weeks. While that provided the business case for purchasing a blue box, the incentive to actually build the blue boxes was entirely different.

Using blue boxes to gain access to the phone system was against the law. The October 1971 edition of *Esquire* magazine spoke of John Draper, one of the preeminent builders of blue boxes, by his code name (Cap'n Crunch). Draper would eventually receive five years' probation for using blue boxes. However, the idea of building a device that would allow placing calls without inserting any coins was intriguing to Wozniak, who had a copy of the magazine article.

> I was so grabbed by the article, I called Steve Jobs before I was halfway through and started reading him passages.
>
> — Steve Wozniak[19]

Jobs and Wozniak thought it was a great story but they were convinced it was just that—a story with no possibility of being true. In order to bring the idea of building the blue box to fruition, they had to have information that was not included in the *Esquire* article.

> And it turned out we were at Stanford Linear Accelerator Centre (SLAC) one night and way in the bowels of their technical library way down at the last bookshelf in the corner bottom rack we found an AT&T technical journal that laid out the whole thing and that's another moment I'll never forget—we saw this journal we though my God it's all real and so we set out to build a device to make these tones.
>
> —Steve Jobs[20]

Wozniak saw building blue boxes as a technical challenge for tinkerers. Jobs saw building blue boxes as a technical challenge for tinkerers—and a means of making a bit of money. Wozniak described part of the technical process that allowed taking over the phone infrastructure at the time, and believed that he would just receive a warning from law enforcement if caught breaking the law while using the blue boxes to impress friends. After a bit of experimentation, the two were able to create reliable blue boxes that always hit the proper tones.

> I was young and knew that I'd just get warned if caught. Also, I was more enamored with having something to amaze people with than in money. I actually made all my own calls on my own dollar. I only used the blue box to pull of hilarious stunts and to explore how far I could possibly get with it. All the blue boxes we sold might have bought a nice hi-fi, which got stolen from my Cupertino apartment anyway. The blue box year was 1972. Apple started in 1975. The biggest connection was some design tricks and techniques that I honed on the blue box.
>
> —Steve Wozniak[21]

As with John Draper, the two decided that making illegal products required each to adopt a code name. Wozniak's was inspired by the university he was be attending, while Jobs named himself either "Oaf Tobar" or "Oaf Tobark."

> Eventually Steve Jobs (a.k.a. Oaf Tobar) and I (a.k.a. Berkeley Blue) joined the group, making and selling our own versions of the Blue Boxes. We actually made some good money at this.
>
> — Steve Wozniak[22]

Even in the early 1970s, a public pay telephone was seen as a harder-to-trace source of prank calls, such as to the Pope.

> "Jobs supplied $40 in parts and sold the boxes door-to-door in dorm rooms for $150, splitting the profits with Wozniak. In keeping with the spirit of 'phone phreaking,' Wozniak assumed the name Berkeley Blue and Jobs, Oaf Tobark. During one demonstration, Wozniak called the Vatican posing as Secretary of State Henry Kissinger and asked to speak to Pope Paul VI."[23]

At a sales price of $150 in 1972, the inflation adjusted price forty years later would have been well above $800—this was not a modest purchase for any student, even the college students at Berkeley at the time. The use of the blue boxes was not without risk; many producers and customers were caught (like Cap'n Crunch). Others were almost caught and/or questioned by police (like Jobs and Wozniak).

> So we're sitting in the payphone trying to make a blue box call. And the operator comes back on the line. And we're all scared and we'd try it again. … And she comes back on the line; we're all scared so we put in money. And then a cop car pulls up. And Steve was shaking, you know, and he got the blue box back into my pocket. I got it—he got it to me because the cop turned to look in the bushes for drugs or something, you know? So I put the box in my pocket. The cop pats me down and says, "What's this?" I said, "It's an electronic music synthesizer."
>
> —Steve Wozniak[24]

As innovators, Jobs, Wozniak, and others learned about the power of technology while tinkering with objects such as blue boxes.

> What we learned was that we could build something ourselves that could control billions of dollars worth of infrastructure in the world—that was what we learnt was that us two you know, we're not much, we could build a little thing that could control a giant thing and that was an incredible lesson. I don't think there would ever have been an Apple computer had there not been blue boxes.
>
> —Steve Jobs[25]

Jobs admitted that the process of developing, testing, and implementing the blue boxes gave the two the confidence to work together on other initiatives. It mattered little to Jobs that they were knowingly breaking the law against "toll fraud," as they were stealing a service from AT&T by not paying the required tolls to make phone calls. To Jobs, the philosophical question was how many laws the pair could be allowed to break in exchange for positive, legal developments at a later point in time.

> There is a societal benefit … to tolerating, perhaps even nurturing … the crazy ones—the misfits, the rebels, the troublemakers, the round pegs in the square holes.
>
> —Phil Lapsley[26]

The references to Jobs's and Wozniak's blue boxes suggest they made money from the initiative, while simultaneously enabling themselves (and others) to withhold revenue from the primary telephone company (AT&T). Wozniak even mentions making enough alone to purchase an expensive stereo system. However, Jobs later attested to the federal government that he made no money from illegally creating blue boxes. As part of the process of seeking a top secret clearance in 1988, he was required to admit committing the crime that was undetected (or at least not prosecuted) in the past.

> I admit in engaging in criminal activity when I discovered one could make long distance phone calls on an electronic device for free. This device was named a "Blue Box." The challenge was not that I could make long distance phone calls for free, but to be able to put a device together than could accomplish that task. I did not make a profit from what I considered this to really be a "project." At the age of approximately fourteen, it was a technical challenge, not a challenge to be able to break the law.
>
> —Steve Jobs[27]

In the disclosure to the federal government, Jobs did not implicate Wozniak in the building of the blue boxes. He also did not reference any profits from the building of blue boxes, despite reports claiming they earned up to $6,000 from the sales.[28] As well, he conveniently told the investigator from the federal government he was approximately 14 years old at the time, but the *Esquire* article detailing the blue boxes came out shortly before his 17th birthday, when he could more easily be perceived as an adult.

HOMESTEAD HIGH SCHOOL

Alongside projects such as building blue boxes, Jobs was still attending high school and would graduate a few months after his 17th birthday. Wozniak later expressed that his fondest memories of Steve Jobs were not of their work with Apple, but of their interactions while Jobs was still in high school. Those interactions included their combined work on blue boxes and other activities, such as the time the two "tried to unfurl a banner depicting a middle-finger salute from the roof of Homestead High School."[29]

Jobs was clearly a profoundly intelligent individual. He even skipped a grade after testing in elementary school. However, he was not the most profoundly exceptional student; the FBI revealed he "enrolled at Homestead High School on September 10, 1968 … attended Homestead High School until June 15, 1972, at which time he graduated. He earned an overall grade point average of 2.65 on a 4.0 scale."[30] One of the individuals who would revolutionize computing at the age of 20 had been a "C" to "C+" student throughout high school, with college yet to come.

Chapter 2

REED COLLEGE AND TIME BEFORE APPLE (1972–1976)

In spending a little time with these people, I noticed some of their other behaviors: they used to sell incense to the local department stores and then go steal it back, so that the department stores would buy more and they would have a thriving business. And their ethics told them that this was fine, that anything in the service of Krishna was fine. In interacting with them I think I learned more about situational ethics than I ever did on campus.

—Steve Jobs, on Hare Krishna

JOBS AS A FAIRY TALE CHARACTER

Good friends Jobs, Wozniak, and Fernandez repeatedly came across each other over the next few years. Immediately after Jobs and Fernandez completed high school, there was a need for some part-time work for spending money over the summer. After Hewlett-Packard, Jobs's next major summer position was dressing up with Wozniak as characters from *Alice in Wonderland* for $3 an hour. This summer position paid about 50 percent more than minimum wage at the time (summer of 1972) but was less risky than the work creating and selling blue boxes.

> To make ends meet in the summer of 1972, Wozniak, Jobs, and Jobs's girlfriend took $3-per-hour jobs at the Westgate Mall in

San Jose, dressing up as *Alice in Wonderland* characters. Jobs and Wozniak alternated as the White Rabbit and the Mad Hatter.[1]

REED COLLEGE

As a condition of adoption, Paul and Clara Jobs made a commitment to his biological mother that Steve would attend college. He briefly attended Reed College in Portland, Oregon. Jobs was enrolled as a degree-seeking student for one semester but was around campus for most of two years.

> He said if he didn't go there he didn't want to go anywhere.
>
> —Paul Jobs[2]

The selection of Reed, a rather expensive university, was a sizable expenditure for his adoptive parents; even more so because "at this stage in his life, Jobs had 'no idea' what he wanted to do and 'no idea' how college was going to help him figure it out."[3] Jobs was attending college partially because of his biological mother's insistence that he should attend.

Dan Kottke, who would become an early Apple employee, remembered meeting Jobs in the first few weeks of their shared first semester at Reed College, bonding over the book *Be Here Now* and the same reel-to-reel music player and music that had previously connected Jobs and Wozniak.

> We met in the first few weeks freshman year at Reed College.
>
> Steve, I remember, had a big Teac reel-to-reel tape deck with lots of bootleg Dylan. It was a pretty expensive piece of gear; I don't know how he wangled that.
>
> —Dan Kottke[4]

The connections Jobs and Kottke made early in their academic experience lasted for many years, but not throughout their entire lifetimes. The experiences from early in that first semester influenced Jobs greatly. At the bookstore, Jobs and Kottke "had both just bought a copy of *Be Here Now* by Ram Dass for $3.33—thus beginning a long friendship that would take them to India and beyond. Student body president Robert Friedland ... was another influential figure in Steve's life. Robert managed a farm in McMinnville (dubbed the All-One Farm), which became a magnet for psychedelic pilgrims—including Daniel and Steve, who spent time at the farm tending the apple trees, an experience that would later inspire the name Apple."[5]

Like Jobs, Friedland was another future billionaire studying at Reed. He had come to Reed College from a lucrative (but failed) illegal business venture that resulted in more punishment than creating blue boxes. He had served time in federal prison after being caught with 24,000 LSD pills, leaving Bowdoin College in 1970 and finishing his studies at Reed after the jail sentence.[6]

> I definitely see that Steve learned a lot from Robert. When we first met him at Reed he had done time in federal prison for being caught with tens of thousands of doses of LSD.
>
> —Dan Kottke[7]

Friedland is also the individual who planted the idea in Jobs's head that he should visit India. Friedland finished his degree but Jobs did not. Although Jobs didn't complete his degree, he continued to attend courses for a year-and-a-half. "He felt guilty about putting his parents to unnecessary expense, so he 'dropped out' after six months at Reed and then 'stayed around,' or 'dropped in,' for 18 more months—auditing classes of his own choosing while paying no tuition."[8]

As Kottke would later explain, Steve Jobs determined he really didn't require college credits but he did require the experience the college courses would provide.

> Steve, using talents that proved valuable later in life, did a cost-benefit analysis and came to the conclusion that he didn't really need the college credit. He withdrew, got all his tuition money back, then went back to the dean and said, "can I just audit some classes?"
>
> —Dan Kottke[9]

Jobs didn't know how his time at Reed would help in the future, but he did know that the knowledge, skills, and abilities there would help, somehow, someday, some way. For instance, the variety of fonts used today on computers was inspired by a single calligraphy course Jobs attended at Reed taught by Father Robert Palladino, an artist rather than a scientist.[10]

> I learned about serif and san serif typefaces, about varying the amount of space between different letter combinations, about what makes great typography great. ... It was beautiful, historical, artistically subtle in a way that science can't capture, and I found it fascinating. None of this had even a hope of any practical application in my life. But 10 years later, when we were designing the first Macintosh computer, it all came back to me. And we designed it all into the Mac.
>
> —Steve Jobs[11]

Officially, Steve Jobs attended Reed College for a single semester. He attended classes for another year-and-a-half but didn't pay for those classes. Jobs mentioned this in his 1991 convocation speech at Reed, titled "Stay Hungry." The "hungry" he mentioned was both literal and metaphorical—Jobs had to struggle to eat during the time he was in Oregon. Each weekend, he walked to get food provided by the Hare Krishna temple, where he got reinforcement for a concept of situational ethics that he would soon show with a later collaboration with Wozniak; it was sometimes fine to be dishonest or steal if it helped the organization meet goals.

> In spending a little time with these people, I noticed some of their other behaviors: they used to sell incense to the local department stores and then go steal it back, so that the department stores would buy more and they would have a thriving business. And their ethics told them that this was fine, that anything in the service of Krishna was fine. In interacting with them I think I learned more about situational ethics than I ever did on campus.
>
> —Steve Jobs[12]

ASHRAM IN INDIA

That summer after our freshman year he (Friedman) went to India and was part of the Be Here Now *scene. He was hanging with the people in that book.*

Somehow I and Steve got to know him. We had an interest in Indian philosophy and spirituality in general. It was Robert who had the idea we should go to India. I had no money. But Steve had found work at Atari so he had money, and we went to see the Kumbh Mela, the biggest religious gathering in the world.

—Dan Kottke

Inspired by Robert Friedland, Jobs and Kottke travelled to India in 1974. They had planned to visit Hindu guru Neem Karoli Baba but there was one major problem. Guru Neem Karoli Baba had died the previous year, which Jobs and Kottke did not yet know.

During the trip, Jobs picked up lice, contracted dysentery and scabies, was trapped in a major thunderstorm while sleeping in a dry creek, got dunked in a lake, and had his head shaved by a beggar. For most individuals, these experiences would be considered dreadful, or at least life-changing in a negative way. For Jobs, his experience was life-changing in a positive way.[13]

During these travels, Jobs met Dr. Larry Brilliant, who would later start the Seva Foundation. When Jobs and Brilliant first met, the "enlightenment-seeking Mr. Jobs showed up with bare feet and a shaved head at the Himalayan ashram where Dr. Brilliant was living."[14] Jobs walked around the area reading *Autobiography of a Yogi*, seeking vegetables for salads.

> He was searching for the same thing all of us search for, what we're still searching for, the meaning of life, why we live, how we can do anything good in our lifetimes.
>
> —Larry Brilliant[15]

According to Brilliant, early in Jobs's stay, "Steve had been flirting with the idea of being *sadhu*." Most Indian *sadhus* live a monk-like existence of deprivation as a way of focusing solely on the spiritual. Jobs's interests migrated toward Buddhism, which allows for more engagement with the world than is permitted ascetic Hindus. Among other things, Buddhism made him feel justified in constantly demanding nothing less than what he deemed to be "perfection" from others, from the products he would create, and from himself.[16]

Walter Isaacson would later relay how Jobs described learning to rely upon intuition rather than intellect during this trip to India.

> He told me he began to appreciate the power of intuition, in contrast to what he called "Western rational thought," when he wandered around India after dropping out of college. "The people in the Indian countryside don't use their intellect like we do," he said. "They use their intuition instead... Intuition is a very powerful thing, more powerful than intellect, in my opinion. That's had a big impact on my work."[17]

Despite the eventful 1970s trip and its many health risks, more than 30 years later, Job recommended Facebook founder Mark Zuckerberg visit the same ashram when he was contemplating the strategic direction of Facebook.

> Everybody in the world wants to go and see this place... It's a combination of *Eat Pray Love*, know thyself and change the world.
>
> —Larry Brilliant[18]

ILLICIT DRUG USE

Starting in high school, Steve Jobs used multiple illegal drugs including marijuana, hashish, and LSD. His drug use was admitted in public sources and was also described in government files. An interviewer described Jobs as stating he believed using LSD was one of the single most important activities he had undertaken, and that no one could fully understand him without having the same experience.

> He explained that he still believed that taking LSD was one of the two or three most important things he had done in his life, and he said he felt that because people he knew well had not tried psychedelics, there were things about him they couldn't understand.[19]

In his own words, Jobs admitted to the government during a Department of Defense background check that he was glad to have experimented with drugs like LSD, although he could not articulate precisely why he was pleased.

> I have no words to explain the effect the LSD had on me, although, I can say it was a positive life changing experience for me and I am glad I went through that experience.
>
> —Steve Jobs[20]

In a part of his FBI file, there are multiple references to his use of marijuana, hashish, and LSD, including a note from one of his classmates at Reed.

> He had not used any illegal drugs in the past five years; however, during the period of approximately 1970–1974 he experimented with marijuana, hashish, and LSD. This was during high school and college and he mostly used these substances by himself.
>
> [Redacted] also advised that he was aware that Mr. Jobs used illegal drugs, including marijuana and LSD, while they were attending college.
>
> Mr. Jobs also commented concerning his past drug use.[21]

Jobs repeatedly insisted that he did not consume any illegal drugs after Apple was incorporated in 1977, as he had multiple interviews with government agencies (the Department of Defense and FBI) for background checks and clearances. In many of these checks, illegal drug use within five years is an automatic disqualifier, and all outside individuals interviewed suggested that Jobs had stopped using illegal drugs in the 1970s.

Despite this, Jobs believed others should share the same experiences he had in India and with illegal drugs. He once commented that Bill Gates may have benefitted most:

> I wish him the best, I really do. I just think he and Microsoft are a bit narrow. He'd be a broader guy if he had dropped acid once or gone off to an ashram when he was younger.
>
> —Steve Jobs

CHRISANN BRENNAN

McKinnell wrote of Chrisann Brennan that she was "the artist Jobs fell in love with when he was a teenager living in his parents' home. She remembers his bright-red electric typewriter and the tidiness of his bedroom closet. The two were high-school sweethearts, dropping acid, living together, converting to Zen Buddhism."[22]Although she and Jobs never married, she would intersect Jobs's life over a period of 40 years. She went from high school sweetheart to Apple employee, and arises in his narrative on numerous occasions in the future.

ATARI AND *BREAKOUT* (THE VIDEO GAME)

Wozniak was older than Jobs, and had started a full-time position within a year of Jobs graduating high school. *Apple Confidential* describes a Wozniak who was happy to be working full-time at Hewlett-Packard; five years earlier, a then-13-year-old Jobs had similarly been happy just to have a summer job at Hewlett-Packard. However, Wozniak had tried to start working on computers instead of calculators.

> On February 20, 1973, Wozniak started working at a "real job" in Hewlett-Packard's Advanced Products Division, and was soon joined by Fernandez, where the two designed new handheld calculators. For Wozniak, it was a dream just to be working at HP, though he'd tried in vain to join the computer division.[23]

After no longer officially attending Reed College, Steve Jobs was able to get a position with video-game–maker Atari in its earliest days. He was hired directly by Nolan Bushnell, the founder of the firm, who realized Jobs was a challenge to deal with but was also frequently the smartest person around, despite his low high school grade-point average and early departure from Reed College.

> And yes, it was at Atari that Bushnell hired Steve Jobs, who was 19 and not a very pleasant fellow. In fact, Bushnell is one of the few people who actually hired Jobs. (Jobs, as CEO of Apple and Pixar and founder of Next, was usually the one doing the hiring.)

> "Steve was difficult but valuable," Bushnell says. "He was very often the smartest guy in the room, and he would let people know that." But Bushnell remains awestruck by Jobs's success and the way he grew into not only a top executive, but a visionary—a rare combination.[24]

Close friend Wozniak described Atari as a good fit for the young Steve Jobs, who wanted the ability to work at times and then spend time in Oregon.

> I must tell you that Steve Jobs was well liked by the Atari execs and he'd work in their plant in Los Gatos for a while, then live up in Oregon with friends for a while, on and off. Games that the Atari engineers designed in Grass Valley would come to Los Gatos and Steve would examine their design and make changes, like adding sounds, etc.
>
> —Steve Wozniak[25]

At Atari, Jobs was tasked to develop the video game *Breakout*, which would soon become very popular in arcades. *Breakout* was a game "... in which the player bounced the ball off a paddle at the bottom of the screen in an attempt to smash the bricks in a wall at the top. Bushnell turned to Jobs to design the circuitry. Initially Jobs tried to do the work himself, but soon realized he was in way over his head and asked Wozniak to bail him out. 'Steve wasn't capable of designing anything that complex. He came to me and said Atari would like a game a described how it would work,' recalls Wozniak. 'There was a catch; I had to do it in four days. In retrospect, I think it was because Steve needed the money to buy into a farm up north.'" [26]

The fundamental problem was that Jobs wasn't capable of completing the assignment; he wasn't talented enough at engineering. Bushnell was aware Jobs had friends such as Wozniak, a gifted engineer, to help on this project. Jobs convinced his friend Wozniak to assist for half of the amount Jobs was to be paid by Atari. Except Jobs didn't tell Wozniak the truth.

In return for completing the game in four days, Jobs promised Wozniak half of the pay from Atari, which would increase based upon the number of microchips he could eliminate from the design. Wozniak received $350 for the design work; which suggested Jobs had split $700 (if he had kept his word to give Wozniak half). The actual payment Steve Jobs received from Atari was $5,000. So Wozniak undertook the majority of design work in the evenings—after working all day at his full-time job with Hewlett-Packard—for $350, while Jobs pocketed his much larger "half" of $4,650.

> I did design Breakout for Atari, in very few chips. Steve got me to do it in 4 days, which is unbelievable for hardware. We both stayed up all night for 4 nights in a row and barely finished it.
>
> —Steve Wozniak[27]

Jobs's deception eventually came out in a book published in the 1980s. Wozniak admitted to crying when he heard about being scammed by his friend.[28] Prior to this, he had been excited just by being presented with the challenge of designing the game, as he routinely built calculators instead.

> I would gladly have designed the BREAKOUT game for Atari for free, just to do it. I had a job at Hewlett Packard. I considered $350 a nice bonus, something that I'd earned myself. I probably had a pizza to celebrate. I was hurt in later years when I heard that Steve was paid more than he'd told me, and I don't think that I hurt easily.
>
> —Steve Wozniak[29]

Wozniak did admit to forgiving Jobs and moving on. He declared that the alternative—not forgiving—would have been worse for him personally. "I'm very forgiving and it would give me a worse feeling had if I kept sadness inside. Better to forgive and forget and remain friends. Good things can come out of that."[30]

However, this was the second time—the first being the Blue Boxes—Jobs had figured out how to use Wozniak's technical expertise in order to make money on an enterprise. At least this time, the business collaboration was not an illegal act.[31]

ARREST IN "APPROXIMATELY" 1975

Only once was Jobs confirmed to have been caught in an illegal act; that act resulted in his arrest. However, the arrest record was not widely known until a file from the Department of Defense was released after his death. Jobs was never caught for using the blue boxes but had been arrested for an unpaid speeding ticket. The only reason the arrest was notable was that Jobs had made a formal declaration stating he had never been arrested.

While seeking top secret clearance in 1988, Jobs was challenged about not reporting the arrest in the mid-1970s. Jobs didn't consider the arrest to be an "actual arrest" (his term), as it was for an unpaid speeding ticket while he was being checked for potential underage drinking. As a result, the government required him to sign an amended declaration that described the circumstances of the arrest.

> In approximately 1975, while being questioned by police behind a store in Eugene, Oregon for being a minor with possible posession [sic] of alcohol, I was arrested for an outstanding warrant for not paying a prior speeding ticket. I did not have any alcohol, therefore was not arrested for that reason. I satisfied the warrant by paying off the fine of approximately $50.00. I did not report this on my Personnel Security Questionaire [sic] (PSQ) because I did not feel it was an actual arrest,it was only for not paying my speeding ticket which I eventually paid.
>
> —Steve Jobs[32]

One could say that the experience of being arrested for an unpaid speeding ticket created a long-term deterrent effect on Jobs. The City of San Francisco revealed in early 2016 that he had a refund due, as he had overpaid at least one parking ticket accrued later in life.[33]

HOMEBREW COMPUTER CLUB

Before the introduction of the Altair in 1975, basic computers cost tens of thousands of dollars, had to be rented by the hour, and were often the size of refrigerators or closets. The Altair wasn't cheap—potential buyers would still have to work a few hundred hours at minimum wage—but it was a way for individuals to begin owning the computer technology available at the time. Jobs and Wozniak were members of the Homebrew Computer Club, a collection of computer hobbyists.

> The clubs were based around a computer kit called the Altair. It was so amazing to all of us that somebody had actually come up with a way to build a computer you could own yourself. That had never been possible. Remember, when we were in high school, neither of us had access to a computer mainframe. We had to drive somewhere and have some large company take a benevolent attitude toward us and let us use the computer. But now, for the first time, you could actually buy a computer. The Altair was a kit that came out around 1975 and sold for less than $400. Even though it was relatively inexpensive, not everyone could afford one. That's how the computer clubs started. People would band together and eventually become a club.
>
> —Steve Jobs

Wozniak had built his own computer model at home. It had no name, and Wozniak wanted to give the blueprints to members of the Homebrew Computer Club. Jobs had a different idea—instead of giving the blueprints away, why not build the computers and sell them? This was the third major collaboration between Jobs and Wozniak, after the blue boxes and Atari's *Breakout*.

Wozniak's kit computer was first shown in 1976 and was for hobbyists who didn't want to take the effort to assemble all the parts for the computer. This was a large step toward computing as we often see today; a computer comes out of the box ready to use. But in this case, the user still needed to find an internal power supply and a display. As noted by Linzmayer, Wozniak, who was just married and still working full-time at Hewlett-Packard, had come up with the computer in his spare time and had never even entertained the idea of selling a computer; that idea came from elsewhere—his friend Steve Jobs.

> By March 1, 1976, less than two months after getting married at the age of 25, Wozniak had completed the basic design of his

> computer, and he proudly showed off his work at the Homebrew Computer Club meetings. Jobs quickly saw the potential to profit from Wozniak's computer, just as he had with Wozniak's blue box. Rather than pass out schematics of the computer for free, Jobs tried to convince Wozniak that they should produce printed circuit boards and sell them as a product. "Steve didn't do one circuit, design, or piece of code," recalls Wozniak. "But it never crossed my mind to sell computers. It was Steve who said, 'Let's hold them up in the air and sell a few.'" Jobs admits, "I was nowhere near as good an engineer as Woz. He was always the better designer."[34]

The computer designed by Wozniak would later be referred to as "Apple." The popular media referring to the earliest computers as the "Apple I" would not happen until after the Apple II was released. A basic minicomputer like the Altair 8800 was still rather expensive as a discretionary purchase at $397. The Apple was even more expensive at $666.66 when initially released. But consider the alternatives. The Altair and Apple could fit on a desk, while computers such as the PDP-10 were the size of a refrigerator and cost almost $20,000. Comparatively, the Altair and Apple were exceptionally affordable.[35]

Further, the Apple products had a design configuration that was not common at the time. While the Altair was designed for a small group of individuals who were interested in how the technology worked (requiring assembly and programming), the Apple products, and especially the Apple II, were designed for the much larger group of individuals who could benefit from using computers without undertaking the process of assembling and then programming.

This innovation 40 years ago was a large step toward the computers of today, where a new computer can be unpacked and running within minutes. Apple also kept the cost of the computer down by recognizing most customers already had a suitable monitor in the form of a television. The ability to use television technology for monitors was patented by Apple less than a week before the Apple II was initially demonstrated to the public.

ALMOST NO APPLE

Despite chancing on the brilliant idea of selling a computer that was largely built in an era when others had to completely assemble one, there was almost no company called "Apple Computer" or "Apple," despite the project designed to sell the Apple. Creating the *company* Apple—rather than the *computer* Apple—was a fortunate circumstance. For Jobs, there was the possibility of seeking another short-term gain (like the blue boxes and *Breakout*) rather than building a long-term business. When Jobs later mentioned that creating a company wasn't in their minds, it's because creating a company truly wasn't in their minds. Jobs originally wanted to sell some computers, then sell the rights to the computer. Jobs and Wozniak went through three different possible sale scenarios.

The initial offer to sell the original Apple was to Atari's Nolan Bushnell (Jobs's employer) to be Atari's first computer product.

> Steve and I demoed it in Al Alcorn's home. But Atari was about to take a giant stride with the first home video game (Pong) and had their hands full and didn't have room for a computer at that time. Then again, we never got to talking a specific deal, like how much money or what royalties or what employment, etc.
>
> —Steve Wozniak[36]

> Then Wozniak went to *his* employer. "I pitched my boss, the calculator lab manager, and got him all excited, but it was obvious it didn't have a place at HP," recalls Wozniak. Although his boss didn't think Wozniak's computer was appropriate for his division, he instructed an HP lawyer to call each division head asking, "You interested in an $800 machine that can run BASIC and hook up to a TV?" Everyone declined, saying, "HP doesn't want to be in that kind of market."[37]

Jobs even offered to sell out to Commodore, before hearing that other entrepreneurs had challenges working with the company. Linzmayer reported that the offer Jobs made was quite aggressive for the time. Wozniak thought that Jobs was asking for too much, despite the time he had put into the project. The offer Jobs made to Commodore was:

> to sell the company for $100,000 in cash, some stock, and salaries of $36,000 a year for himself and Wozniak.
>
> "I thought it was atrocious. I had put a man-year of work into it, and I thought it was grossly outrageous to ask for so much," Wozniak naively recalls. Nonetheless, he would have gladly taken the deal because his passion was building computers, not companies. His father was also appalled by Jobs's demands, but not for the same reasons. He felt that Jobs was taking advantage of his son. "You didn't do [*expletive*]," Jerry Wozniak told Jobs.[38]

While Jobs became known as an exceptional businessperson, this would be just the *first* revolutionary business concept he attempted to sell on at least three separate occasions before finding major success. And unlike Apple (which was never sold), Jobs would be successful in selling the next two businesses he created.

Part 2

FIRST RUN AT APPLE

I feel incredibly lucky to be at exactly the right place in Silicon Valley, at exactly the right time historically where this invention has, has taken form.

—Steve Jobs[1]

I just wanted to be in engineering only – I never wanted to run a company, never wanted to run things, step on other people – Steve very clearly did, and wanted to be a top executive and a really important thinker in the world.

—Steve Wozniak[2]

Both Jobs and Wozniak readily admitted that they were both at the right place at the right time. Wozniak only wanted to do engineering, and his products—blue boxes, Atari's *Breakout*, and Apple—would not have been distributed without Jobs. As demonstrated on Atari's *Breakout*, Wozniak's technical design skills were far beyond the capability of Jobs. Neither Wozniak nor Jobs could have created Apple on their own, so each was indeed incredibly lucky.

Chapter 3

THE FIRST YEAR

Without the Homebrew Computer Club, Jobs and Wozniak would not have had the ability to gauge the market's interest in their product. Participation in the Homebrew Computer Club was also a peculiarity of time and place; there were few other places in the world—in 1975—where this could have occurred.

BORN AT THE RIGHT TIME

Malcolm Gladwell suggests that many individuals, such as Steve Jobs and Bill Gates, were born at precisely the right time in history to lead some form of major social or business revolution. In the case of Jobs and Gates, it would be the computer revolution. Steve Jobs and Steve Wozniak already had experience with Atari, Hewlett-Packard, and with building their own electronic devices on their own time, all at an exceptionally young age. Wozniak even had the benefit of being allowed to take electronic components from work when tinkering on his own projects.

At the time the story of the Altair 8800—the computer coveted by the Homebrew Computer Club—was published in late December 1974 (although listed as the January 1975 issue of *Popular Electronics*), the names that became preeminent in computing were all very young, mostly under the age of 20. For instance, Steve Jobs and Bill Gates were both 19, each born in 1955: Steve Jobs on February 24 and Bill Gates on October 28.

The founders of other technology companies were also similarly aged—Scott McNealy of Sun Microsystems was born in November 1954. Nathan Myhrvold wrote of the time period:

> If you're too old in nineteen seventy-five, then you'd already have a job at IBM out of college, and once people started at IBM, they had a real hard time making the transition to the new world. . . . You had this multibillion-dollar company making mainframes, and if you were part of that, you'd think, Why screw around with these little pathetic computers?
>
> —Nathan Myhrvold[3]

If Steve Jobs was a decade older and with HP or IBM, there's a substantial probability Apple would not exist; Steve Wozniak admittedly had no desire to start a company and was working on calculators, not computers. IBM was the biggest name in computing at the time, and frequently posed threats to the firms who started in their shadow (sometimes the threats were real and sometimes perceived).

Steve Jobs was heavily committed to the Apple project, while Wozniak didn't necessarily see his involvement in Apple as requiring full-time effort—in fact, he had developed both the original Apple (retroactively called the Apple I) and the Apple II while working full-time at Hewlett-Packard. Wozniak switched to full-time work at Apple because early investors required him to do so; Wozniak has stated he would have actually been happy to remain at HP.

Among early Apple employees, like Jobs's and Wozniak's friend Bill Fernandez, it was very clear that the opportunity at Apple was driven by the lack of employer entrenchment Myhrvold mentioned. As he would later state in an interview with *Tech Republic* about his decision to start working at Apple:

> These are a couple of my friends, and not corporate types with a lot of stability, and I'll be working in a garage. But, I'm living from home, and I'm not married.
>
> —Bill Fernandez[4]

Being born at the right time alone was not sufficient. Being in the right place and right time was not sufficient. Few technology companies in the history of computing have lasted four decades. Yoffie and Cusumano looked for connections between three of the notable leaders who have built, maintained, and/or resurrected companies over that timeframe—Bill Gates (Microsoft), Andy Grove (Intel), and Steve Jobs (Apple)—and found that

> Gates, Grove, and Jobs owed part of their success to the explosion of activity launched by the invention of the personal computer, the advent of the Internet, and the widespread adoption of digital mobile devices. They were undoubtedly in the right place at the right time. However, many well-positioned businesses run by talented and hard-working individuals failed or fell behind during this very same period and in the same markets. Gates, Grove, and Jobs stand out because they achieved and maintained dominance in their industries, even as seismic shifts altered the landscape around them.[5]

THE HOUSE (AND GARAGE)

The house (and garage) of Apple's founding is located at 2066 Crist Drive, Los Altos, California 94024. It is still owned by Patricia Jobs, Steve's adopted sister. The house number in the 1970s was 11161 Crist Drive. Details about the house were disputed when the city of Los Altos was adding the home to the list of historic places (over the objections of property owner Patricia Jobs).[6]

> Steve Jobs built the first 100 Apple 1 computers at the Crist Drive home with help from Apple co-founder Steve Wozniak and Patricia Jobs. The first 50 were sold to Paul Terrell's Byte Shop in Mountain View for $500 each, according to the evaluation. The rest were assembled for their friends in the Homebrew Computer Club.[7]
>
> Jobs said the evaluation contains inaccuracies about her brother and the founding of Apple Computer Co. For instance, she said Paul and Clara Jobs moved the family from Mountain View to Los Altos in 1967, not 1968, and that the company was conceived prior to 1976. She recalled helping build circuit boards in the living room of the home in 1975.[8]

However, the concept of the company being founded in the garage was largely a myth. Despite popular misconceptions, Apple wasn't founded in the garage—the "company" was originally in Jobs's bedroom, until space became a concern. And even when the "company" had moved to the garage, Wozniak was still building the computers in his own apartment or office. In 1976, there was no concern about an HP employee building a computer for Apple—that would be a clear conflict of interest and conflict of commitment today.

> We didn't build the computer in a garage ... I built most of it in my apartment and in my office at Hewlett-Packard, where I was working at the time. We just used the garage to assemble the parts toward the end. I don't know where the whole garage thing came from. Maybe it's because Bill Hewlett and David Packard built their machine in a garage, everyone assumed we built ours there, too. But really, very little work was done there.
>
> —Steve Wozniak[9]

The naming of the computer Apple came during this early timeframe. Sklarewitz relayed as early as 1979 that "Jobs, who was a California 'fruitarian' at the time the enterprise was moving from bedroom to garage, liked the simple, crisp, natural image of an apple." The founding of Apple was also shortly after Jobs's visits to Friedman's orchard; there were also orchards near the Jobs family home.[10]

THE THIRD CO-FOUNDER

As with the idea of the garage, the idea that Jobs and Wozniak were the sole co-founders of Apple is also a myth. On April 1, 1976 (April Fools' Day), Apple Computer Company was founded by Steve Jobs, Ronald Wayne, and Steve Wozniak in the Jobs family home. Ronald Wayne?

Everyone knows Apple had two primary founders—Steve Jobs and Steve Wozniak. Except there were actually three; the third was Ron Wayne, one of Jobs's co-workers at Atari and the third co-founder of the Apple partnership while the co-founders were engaged in full-time work elsewhere. Wozniak wrote about Wayne:

> Ron Wayne was a third partner for the Apple I 'side' business we started. Steve had 45%, I had 45%, and Ron had 10%.

> He wrote manuals and could decide things if Steve and I differed.
>
> —Steve Wozniak[11]

Except Wayne was a partner in Apple for a very short period of time, less than two weeks total. He realized one of the biggest problems with partnerships was unlimited liability. Older than both Jobs and Wozniak, he was more established in his career and had assets—Jobs and Wozniak had to sell a vehicle and expensive calculator just to have some funds to start the business.

Under the concept of unlimited liability in a partnership, every individual partner is responsible for the entire debt of the business. If the company had placed a large order for microchips and their Apple computer didn't sell, Wayne would be responsible for paying all the bills due to creditors because his partners simply didn't have any funds for the creditors to claim.

> "Either I was going to be bankrupt or the richest man in the cemetery," Wayne recalls thinking. Because Apple was far from a sure thing, Wayne retained his day job at Atari and worked nights writing documentation and designing a logo for the infant company. The logo he created was a pen-and-ink drawing of Sir Isaac Newton leaning against an apple tree with a portion of a William Wordsworth poem (*Prelude, Book III, Residence at Cambridge*) running around the border: Newton. ... A mind forever voyaging through strange seas of though ... alone."
>
> —*Apple Confidential* [12]

For the sum of $2,300, Wayne gave up his 10 percent of the firm. Leaving the partnership so early, the minimal buyout from Apple in 1976 and 1977, plus the modest funds he made from selling his copies of the partnership agreement signed by he, Jobs, and Wozniak to a collector years later are the only compensation Wayne ever received from his 11 days as a co-founder of Apple.[13] Four decades after the founding of Apple, Wayne would declare:

> You know, I've never been very successful at business. I've never been rich. But I've never been hungry either. And I can say

> this: I created something with my own hands that actually sold for $1 million.
>
> —Ron Wayne[14]

The three original founders of Apple would not meet again until more than 20 years later. As detailed in *Apple Confidential*:

> Out of the blue in August 1999, Jobs invited Wayne to meet him in San Francisco. The two toured the Seybold Seminar at the Moscone Center in the morning, then met again for an afternoon luncheon on the patio at the Apple cafeteria in Cupertino where they were joined by Wozniak. The trio spent a pleasant hour and half catching up and reminiscing. It is the first and only time the three founders of Apple have reunited since parting ways in 1976.
>
> — *Apple Confidential* [15]

APPLE COMPANY

> *When we started Apple, we were out to build computers for our friends. That was all. No idea of a company.*
>
> —Steve Jobs[16]

Jobs readily admitted the initial purpose for Apple was to sell some computers, rather than building a permanent business that could sell millions of units. Building the computers allowed Jobs and Wozniak to make money in addition to their full-time work, and they had some orders pre-committed from the Homebrew Computer Club. And each party was doing what individually brought them joy.

> [Jobs] wanted to do things, [whereas] I was more the engineer-technologist that wanted to build things. We had different wants in the end.
>
> —Steve Wozniak[17]

> After Steve met me, he never tried to be the designer of the pair. … He always thought in terms of products: how are they going to

> affect people? It's not how you connect a few chips together, it's what are they going to do that's useful. That's ... a marketing point: you have to think of the end user. And that should always be number one.
>
> —Steve Wozniak[18]

> Jobs was a marketing guy who often thought more like an engineer. His answer to an open-ended problem wasn't merely an answer, it was the answer. Woz is an engineer who can see more than one answer to even the most well-defined engineering challenge. In his memoir, iWoz, he makes the case that the best engineers—the ones who think beyond the "artificial limits everyone else has already set"—are artists who "live in the gray-scale world, not the black-and-white one.
>
> —Jon Zilber[19]

> And Steve to his credit was harnessing Woz. Woz did not have the entrepreneurial urge and was really happy at his job at Hewlett Packard, and I think he's been on record as saying he would have been happy staying there.
>
> —Dan Kottke[20]

As Apple began to sell computers, Wozniak immediately deferred to Steve Jobs, taking the nickname "Woz" to eliminate the confusion of an operation comprised of two co-founders each named "Steve."

> I only started being called Woz as we started Apple, as there were two Steves.
>
> —Steve Wozniak[21]

The first Apple computers created for retail were sold to Paul Terrell's "The Byte Shop" at $500 each. Jobs and Woz worked together to derive what would be listed as the Manufacturer's Suggested Retail Price (MSRP) of the Apple. The price the pair agreed upon for retail purposes was $666.66, which some potential consumers thought had a religious implication (whether intentional or inadvertent). The suggested sales price was instead completely accidental, based upon Woz liking numbers that repeated.

> I've always collected good phone numbers with repeating digits ... We first knew that the Byte Shop of Palo Alto would be buying the Apple I computers from us for $500 each. Steve came up with something like $650 or $666 or, quite likely, $667, which is a logical 4:3 ratio. I spotted this and added .66 to keep all the digits the same. I'd never attended church nor read the bible so I didn't know of any negative connotation with 666.
>
> —Steve Wozniak[22]

From the moment Jobs declared the Apple should be sold as an assembled product instead of giving away the blueprints, Apple had created a computer that was palatable to both the members of the Homebrew Computer Club, who didn't want to assemble a computer, and members of the general public, who couldn't assemble a computer.

> The customers for the Apple I were Woz and me and our friends in the Homebrew Computer Club. The Apple I was really the first computer to address the needs of the hobbyist who wanted to play with software but could not build his or her own hardware. It came with a digital circuit board, but you still had to go get your own keyboard, power supply, and television monitor. If you were a techie, the Apple I seemed to go 90 percent of the way. Of course, if you weren't a techie, it only went 10 percent of the way. We sold almost 200 of the Apple I.
>
> —Steve Jobs[23]

> Now we made the, a very important decision was to not offer our computers a kit. Even though you needed to buy these extra parts. The main computer board itself came fully assembled. We were the first company in the world to do that. Everybody else was offering their little computers a kit. And what that meant was there was maybe an order of magnitude of more people who could actually buy our computer and use it than if they had to build it themselves.
>
> —Steve Jobs[24]

The original Apple computer put money into the pockets of more than one Jobs family member; Jobs's adopted sister Patricia was paid to assemble the boards by inserting chips. She was admonished for bending pins, as that made the chips unusable.

> We'd pay Patti Jobs and other friends $1 per board to insert all the chips from boxes of chips that we had.
>
> —Steve Wozniak[25]

> I'd get yelled at if I bent a prong.
>
> —Patricia Jobs[26]

Patricia Jobs received a better deal than Jobs's college friend Dan Kottke; he was paid $3 per hour but could assemble far more than three boards per hour.

PERSONAL COMPUTING 1976

Jobs, Wozniak, and Kottke attended the Personal Computing 1976 conference in Atlantic City, New Jersey, when only the 200 original Apple computers were created in the entire history of Apple. Jobs and Kottke were the public faces of the firm, interacting with potential customers at the booth, while Wozniak stayed in the room working on the BASIC programming language. For Wozniak, it was the first time he had left the State of California (except for the brief time he spent at school in Colorado and trips to Tijuana, across the U.S./Mexico border from San Diego).

> PC '76 was very important to me. It was only my second time out of California, the first being a year of college in Colorado. I don't count Tijuana as being out of California. At PC '76 I sat in our room upstairs and wrote additions to my BASIC while Steve Jobs manned the floor in the daytime.
>
> —Steve Wozniak[27]

Even at this early event, Jobs and Wozniak were each doing what made them happiest. Jobs was the public face talking about the power of computing and making sales, while Wozniak was working on solving problems with a programming language that would open new possibilities with the computer he had created.

Chapter 4

APPLE COMPUTER INCORPORATED AND GROWTH

REVOLUTIONARY APPLICATION

Wozniak had made the Apple and Apple II, plus he had also written programming languages like BASIC. Apple still needed a major software application in order to move computers from consideration as a toy for hobbyists to a tool used in homes and businesses. Jobs would later describe the event he considered to be the launching point for Apple (the corporation), the first spreadsheet (VisiCalc).

> The first one … really happened in 1977. And it was the spreadsheets. I remember when … Dan Fylstra who ran the company that marketed the first spreadsheet, walked … my office at Apple one day and pulled out this disk from his vest pocket and said, "I … I have this incredible new program. I call it a visual calculator." And it became VisiCalc. And that's what really drove, propelled the Apple to … to the success it achieved more than any other single event.
>
> —Steve Jobs[1]

INTRODUCING APPLE II

Steve Wozniak was the technical genius—both hardware and software—of the early Apple. He holds a total of four patents, including one filed on Monday April 11, 1977 entitled "Microcomputer for Use with Video Display." The patent was assigned to Apple Computer, Inc., which meant Apple—the corporation—owned the intellectual property, not Jobs or Wozniak. And patents assigned to corporations remain with the corporation, even if the inventor leaves.[2]

The Apple II computer was introduced the following Saturday, on April 16, 1977, at the West Coast Computer Faire, with Apple owning the patent on using television sets as a monitor for computers.

1977—WEST COAST COMPUTER FAIRE

At the West Coast Computer Faire, Jobs and Wozniak both knew that the Apple II was a major product (despite small numbers of original Apple computers created and sold); they paid for the best booth and rented a video projector when projectors were not yet a common technology.

> Steve Jobs got the info packet. Both he and I felt that we had such a good product that we should immediately secure the prime booth spot, which we did. We also arranged to rent a video projector. This was such an early year that such projectors were virtually unknown. It was a BIG deal.
>
> —Steve Wozniak[3]

While at these events, having the best possible booth may have been an unnecessary over-commitment at the time. Apple was providing the only real opportunity for someone wishing to use a computer rapidly, as even the power supply was now integrated. Other options required assembly and extensive programming just to find minimal uses; customers would find the Apple.

ROLES OF THE FOUNDERS

Wozniak frequently talked about how Jobs wasn't an engineer, designer, or coder but was proficient enough to make revisions if provided with a functional draft.

> Steve didn't ever code. He wasn't an engineer and he didn't do any original design, but he was technical enough to alter and change and add to other designs.
>
> —Steve Wozniak[4]

> … he was a good point for discussion of higher level things. He could understand the designs and code to an extent (well, like processor data pins going to the RAM data pins) but wasn't capable of matching me so he didn't try.
>
> —Steve Wozniak[5]

But Jobs wouldn't let his absence of formal training in engineering, design, or code stop him; to Jobs, his answer was almost always the answer:

> Jobs was a marketing guy who often thought more like an engineer. His answer to an open-ended problem wasn't merely an answer, it was the answer. Woz is an engineer who can see more than one answer to even the most well-defined engineering challenge. In his memoir, *iWoz*, he makes the case that the best engineers—the ones who think beyond the "artificial limits everyone else has already set"—are artists who "live in the gray-scale world, not the black-and-white one."
>
> —Jon Zilber[6]

For instance, Jobs repeatedly challenged fundamental principles of nature and physics to meet his standards. The first design of a color Apple logo had the colors presented in the order of the rainbow. Jobs liked the idea of using the colors of the rainbow, but decided to change the order of the colors:

> Steve and I wanted a color logo with the Apple theme and the Regis McKinna agency, which we were just hiring, came up with it. Steve Jobs rearranged the colors with the darker blue at the bottom and lighter green at the top, rather than keeping them in rainbow order.
>
> —Steve Wozniak[7]

While ROYGBIV is a common pneumonic device for remembering the colors of the rainbow (red, orange, yellow, green, blue, indigo, and violet), the original color apple logo was ordered BVROYG (blue, violet, red, orange, yellow, green), from bottom to top.

MIKE MARKKULA—THE FIRST INVESTOR

Mike Markkula, a quasi-retired millionaire from Intel, provided the initial funding to help place Apple on a growth path with the introduction of the Apple II. The investment was truly needed, as Markkula knew the company had to grow rapidly to have any chance against competitors.

> Jobs and Wozniak were young and bright, and also mature. They realized they had zero experience in management and if Apple was to grow as fast as we projected, we'd need able people with experience in big companies. ... It was go for broke ... We had to dominate the business or go bankrupt trying.
>
> —Mike Markkula[8]

Early employee Mark Johnson described that the company went from a little startup to an assembly line in a very short period of time. Jobs was already a demanding individual and Johnson's mother was the reason he had gotten his job at the company.

> In the short time I was there it went from a goofy startup to a full-blown production line ... my mom had this relationship with Steve, Steve liked my mom ... He had an interview on TV in Oakland, and I got drafted to drive him in my VW bug. Steve, even at that point, he was a handful, an intense person.
>
> —Mark Johnson[9]

Markkula knew in early 1977 that his primary competition would include IBM, even though he mistakenly believed that Texas Instruments would also be a big threat to Apple. The larger firms could use earnings from other products to cover expenses that were comparatively large for a start-up like Apple. The case design for the Apple II alone did not come cheap—it cost $100,000 just for the machinery.

Ramping up production would require another investor not yet identified by Markkula.

Markkula also understood he had to impose some discipline on Apple to always get what the company wanted and needed. In an environment where a lot of competitors were soon going to request computer parts or space in the warehouses of distributors, he made sure every vendor was paid on time. Even in the earliest days, this put Apple at the front of the queue, ahead of competitors when it came to limited resources.

> There have been times when we were the only company able to ship because we could get the parts that others couldn't.
>
> —Mike Markkula[10]

ARTHUR ROCK—THE VENTURE CAPITALIST

But we needed some money for tooling the case, things like that, we needed a few hundred thousand dollars.

—Steve Jobs

Markkula's idea for a venture capitalist was Arthur Rock, who had helped create other successful technology firms. As a venture capitalist, Rock would potentially own part of Apple but wouldn't be as involved on a day-to-day basis as Markkula. Rock was not quite certain that Apple—specifically Jobs—was worth the investment.

> **Arthur Rock:** Well, he wore sandals and he had long, very long hair and a beard and a moustache, but very articulate. He was, I think at one time in his life, and it was probably when I first met him that he ate nothing but fruit.
>
> **Bob Cringely:** So as a mainline venture capitalist, is this …
>
> **Arthur Rock:** This is not the norm. This is not the norm.[11]

He was shown an early Apple, then went to the West Coast Computer Faire. He was initially unimpressed, walking past empty booth after empty booth before finding a crowd of people.

> Then I got to the Apple booth and I couldn't get close to it. I couldn't even get close enough to see the products. People were

> lined up, trying to get into this booth and look at the products. I just figured to myself, there's got to be something here.
>
> —Arthur Rock[12]

Rock figured that some people would spend a few thousand dollars for a computer but thought Jobs was overestimating the number of potential users; Rock invested expecting a positive return but not expecting that Apple—or any computer—would become a standard household appliance. He wasn't particularly impressed with Jobs or Wozniak, especially in relation to Jobs's personal hygiene in the 1970s.

> Then I met Steve Jobs and Steve Wozniak. They kind of turned me off as people. Steve had a beard and goatee, didn't wear shoes, wore terrible clothes, hair down to his collar, and probably hadn't had a haircut in twenty years. But because Mike was so interested, and he had by that time bought a third of the company, I decided that I'd make an investment. And it turned out to be a pretty good investment. Steve's rhetoric, and he had plenty of rhetoric, was that everybody would use a personal computer eventually. I didn't think it would be quite as big a market as it turned out to be. But I could see people spending $2,000 and $3,000 to buy computer for their own use.
>
> —Arthur Rock[13]

The investments from Markkula and Rock allowed Apple to begin production and become the most common computer for educational and home use in a very rapid timeframe.

MICHAEL SCOTT—APPLE'S FIRST CEO

Steve Jobs had a legitimate claim that he was employee number 1 at Apple, but the first CEO of the firm would intentionally ensure that Jobs didn't get the bragging rights from possessing that employee number.

> One of the first things was that of course, each Steve wanted number 1. I know I didn't give it to Jobs because I thought that

> would be too much. I don't remember if it was Woz or Markkula that got number 1, but it didn't go to Jobs because I had enough problems anyway.
>
> —Michael Scott, Apple's First CEO, on assigning Steve Jobs Employee Number "2"[14]

Steve Jobs was not the first CEO of Apple. In fact, he wasn't even among the first five Apple CEOs. The first CEO of Apple was Michael Scott, who needed to give each employee a number for the company's payroll to be processed accurately. He intentionally gave Jobs number 2, as he thought providing Jobs with the symbolic employee number 1 would create additional management problems.

Although Scott was the first CEO of Apple, the firm was no longer headquartered in the Jobs family home.

> I never got to see the garage, I just saw it at Markkula's place up on a hill. Jobs did the talking, and Woz was the quiet one, although more lately Woz has found his voice more. In the early days, we were all so busy, that it was well partitioned over who did what. Woz was doing circuit board itself, Jobs was handling rest of Apple II, Markkula was working on marketing, and I was working on getting us into the manufacturing and all the rest of the business parts.
>
> —Michael Scott[15]

Scott was very aware of Jobs's demanding nature when it came to colors and design; not only with inventing new colors but spending well over a month's timeframe discussing a minor concept such as how rounded the edges of the Apple II case should be.

> The Apple II case came, it had a beige and a green, so for all the standard colors of beige available in the world, of which there are thousands, none was exactly proper for him. So we actually had to create "Apple beige" and get that registered.
>
> I stayed out of it but for weeks, maybe almost six weeks, the original Apple II case, Jobs wanted a rounded edge on it so it didn't have a hard feel. They spent weeks and weeks arguing

> exactly how rounded it would be. So that attention to detail is what Steve is known for, but it also is his weakness because he pays attention to the detail of the product, but not to the people.
>
> —Michael Scott[16]

And just as Jobs understood the original Apple would help hobbyists who needed a computer that was 90 percent assembled and ready-to-go, he understood that the Apple II would reach the mass audience who needed a 100 percent ready-to-go computer to immediately start working with software.

> The Apple II had a few qualities about it. Number one, it was the first computer ever with a plastic case on it. You could mold it and shape it to be a more cultural shape rather than just a rectangular box. And secondly, it was the first personal computer with color graphics on it. Third, in everything it did, it was the first PC that came fully assembled. Every other computer came in a kit. We figured for every hardware hobbyist out there, there was at least a thousand software hobbyists. People who'd want to play with the software but couldn't build one. Even back then, that was how we were thinking.
>
> —Steve Jobs[17]

Scott also described problems at leadership meetings, where he had to preside over a chain-smoking Chairman of the Board (Markkula) and a co-founder (Jobs) that saw nothing amiss with placing his dirty feet and sandals on the conference table.

> The other argument at the meetings was would Steve take his dirty feet and sandals off the table, because he sat at one end of the conference table, and Markkula sat at the other end chain smoking. ... I had the smokers on one side and the people with dirty feet on the other.
>
> —Michael Scott[18]

Jobs was also demanding on the business side, occassionally interjecting himself into contract negotiations where he wasn't really needed.

Even in the early days of Apple, Jobs had significant power over his Chairman and CEO, whether the audience was customers, vendors, or the media. He was the public face of a company despite not possessing either of the highest leadership roles.

> If we were negotiating price for parts, we could negotiate a price with a vendor and at the last minute, Steve would come in and bang on the table and demand to get one more penny off. And of course they would give him one more penny off. Then he'd crow "well I see you didn't do as good a job as you could've getting the price down."
>
> And I'm saying, "Yeah but that one more penny might've cost us a bit more ill will for times when parts are in short supply."
>
> —Michael Scott[19]

Steve Jobs's official title in the late 1970s was "Vice President, Operations," as evidenced on letters he would send to customers offering to take back original Apple computers in exchange for large discounts on Apple II computers. Wozniak was the only person who understood how the original Apple worked, so the company offered to buy all of them back to free up Wozniak's time for work on the Apple II series.

In 1979, two years after the founding of the corporation, the company was already projecting more than $100 million in annual revenue, which was soon to make Jobs (and many others) exceptionally wealthy.

> At the time the company was incorporated, in January 1977, its sales were negligible. By the end of that first year, sales totaled less than $2 million. In 1978, sales were about $15 million. And in 1979, Apple expects to top $100 million.[20]

A WALK IN THE PARC (XEROX PALO ALTO RESEARCH CENTER)

Xerox's PARC (Palo Alto Research Center) had been founded in 1970 based upon Xerox's belief of a paperless office coming by the year 1990.[21] Although Xerox had an incorrect idea of the future, its work at PARC greatly influenced the industry. Xerox—which had started as a photographic company but made its biggest impact on the world through the introduction of the photocopier—was greatly worried about a world where paper was no longer needed. Later, Xerox would influence Apple and Microsoft into developing a lot of the innovations still used in computers today, such as graphical displays and computer mice or pointers.[22]

The cost of Jobs and other Apple personnel visiting Xerox PARC has been disputed. The tours were either free or in shares of Apple stock issued before the IPO; no money was ever exchanged between Apple and Xerox for visits that would change computing.

> In exchange for $1 million of pre-IPO stock, Xerox gave Apple access to its PARC facilities, where Jobs and others saw the progress Xerox was making with the GUI. That visit led to the Apple Lisa—a forerunner of the Macintosh that sold for nearly $10,000 and was never a success—and then the Mac.[23]

> My understanding is not the most direct but I felt that Xerox was letting us in for free.
>
> —Steve Wozniak[24]

Jobs later recalled that more than one of his friends said he truly needed to pay a visit to PARC:

> When I was at Apple, a few of my acquaintances said, "You really need to go over to Xerox PARC (which was Palo Alto Research Center) and see what they've got going over there." They didn't usually let too many people in but I was able to get in there and see what they were doing. I saw their early computer called the Alto which was a phenomenal computer and they actually showed me three things there that they had working in 1976. I saw them in 1979. Things that took really until a few years ago for us to fully recreate, for the industry to fully recreate in this case with NeXTStep. ... And the three things were graphical user interfaces, object oriented computing and networking.
>
> —Steve Jobs [25]

PARC had fantastic ideas but wasn't thinking about what could be done in the mid-1970s—they were thinking about the year 1990 and the paperless office. Further, Xerox had never made a computer intended for sale.

> The problem was that Xerox had never made a commercial computer. This group of people at Xerox was ... more concerned with ... looking out fifteen years than they were looking out fifteen months trying to make a product that somebody could use. So there were a lot of issues that they hadn't solved like menus, other things like that. And at Apple what we had to do was to do two things. One was complete the research which really was only about fifty percent complete. And the second was to find a way to implement it at a low enough cost where people would buy it. And that was really our challenge.
>
> —Steve Jobs[26]

A few years later, Jobs was much harsher in his assessment of Xerox's inability and ineptitude in creating a commercial product, turning a potential history-changing innovation into a non-factor in terms of the computing industry. He suggested that Xerox could have even supplanted IBM if the firm acted upon what the PARC staff had developed.

> Basically they were copier heads that just had no clue about a computer or what it could do. And so they just grabbed ... grabbed defeat from the greatest victory in the computer industry. Xerox could have owned the entire computer industry today. Could have been you know a company ten times its size. Could have been IBM - could have been the IBM of the nineties. Could have been the Microsoft of the nineties.
>
> —Steve Jobs[27]

Fifteen years after his first visit to PARC, Jobs spoke about his frustrations given the amount of time that was needed to make the innovations he had first seen in 1979 commonplace through the world.

> People say sometimes, "You work in the fastest-moving industry in the world." I don't feel that way. I think I work in one of the

slowest. It seems to take forever to get anything done. All of the graphical-user interface stuff that we did with the Macintosh was pioneered at Xerox PARC [the company's legendary Palo Alto Research Center] and with Doug Engelbart at SRI [a future-oriented think tank at Stanford] in the mid-'70s. And here we are, just about the mid-'90s, and it's kind of commonplace now. But it's about a 10-to-20-year lag. That's a long time. The reason for that is, it seems to take a very unique combination of technology, talent, business and marketing and luck to make significant change in our industry. It hasn't happened that often.

—Steve Jobs[28]

WHOSE IDEA WAS IT ANYWAY?

Jobs maintained for the rest of his life that Microsoft—a former Apple supplier—had stolen Xerox PARC's innovations from Apple. Microsoft insisted that both companies (Apple and Microsoft) had each independently stolen the innovations from Xerox PARC; Bill Gates had also visited the facility. As told in *American Spectator* and by Walter Isaacson:

> Jobs worried and raged that Microsoft as an Apple supplier had been stealing Apple's pioneering and user-friendly technology in this area. Gates countered that Apple had, only a few years earlier, copied the same technology from Xerox PARC. That set the stage for a classic confrontation soon after Gates revealed that he would develop a new operating system for IBM PCs featuring a new point-and-click navigation system (much like Macintosh, introduced two years earlier) that would be called Windows. As Isaacson tells the story:
>
> Gates found himself surrounded by ten Apple employees who were eager to watch their boss assail him. Jobs didn't disappoint his troops. "You're ripping us off!" he shouted. "I trusted you, and now you're stealing from us!" [Andy] Hertzfeld [one of Jobs's lieutenants] recalled that Gates just sat there coolly, looking Steve in the eye, before hurling back what became a classic zinger. "Well, Steve, I think there's more than one way of looking at it. I think it's more like we both had this rich neighbor named Xerox and I broke into his house to steal the TV set and found out that you had already stolen it."[29]

Apple eventually sued Microsoft for "stealing" their innovations. Xerox in turn sued Apple, as the innovations were in fact generated by Xerox and the firm didn't want Apple to establish legal rights to ideas that were actually created by Xerox (not Apple). The lawsuits cost millions of dollars and were eventually dropped completely. A few years after the legal debacle, Apple employee Guy Kawasaki succinctly reiterated Jobs's initial comment that Xerox would not be capable of bringing their innovations to market.

> Xerox is incapable of turning a vision into a product. Xerox can't even sue you on time.
>
> —Guy Kawasaki of Apple[30]

UNEASY PARTNERSHIP WITH MICROSOFT

> *I don't think that Steve Jobs had that great of respect for Bill Gates in the early days.*
>
> —Steve Wozniak[31]

In the late 1970s, Apple Computer Incorporated was already shipping Apple II computers that were found in most schools and some homes. Microsoft was focused on creating software for any and every computer manufacturer that wanted software, including Apple. While Apple and Microsoft were very different business in the 1970s, Steve Jobs would later express admiration for what Gates had done with building a software company. Partially due to problems in the protection of intellectual property at the time (there were no penalties or sanctions for pirating software), few software companies existed before Microsoft.

> Bill built the first software company in the industry and I think he built the first software company before anybody really in our industry knew what a software company was, except for these guys.
>
> —Steve Jobs[32]

Bill Gates similarly expressed admiration to Jobs for seeing that there was a broad market of prospective computer purchasers for ready-to-go machines and a comparatively tiny market for those who were able to (and willing to) build their own computers.

> What Steve's done is quite phenomenal, and if you look back to 1977, that Apple II computer, the idea that it would be a mass-market machine, you know, the bet that was made there by Apple uniquely—there were other people with products, but the idea that this could be an incredible empowering phenomenon, Apple pursued that dream.
>
> —Bill Gates[33]

CHRISANN BRENNAN AND JOBS'S FIRST CHILD

Brennan was the mother of Jobs's first child, Lisa Brennan (later Lisa Brennan-Jobs), born in 1978, an early year in the history of Apple. At the time, the Apple II computer was demonstrating rapid sales growth. Chrisann Brennan later released a memoir—after Jobs refused to give her money—that described her interaction with Jobs. She included information about the early days of Apple, when Jobs wanted college friend (and Apple employee) Dan Kottke to live with them.

> It was around this time that Steve, Daniel, and I moved into a rental in Cupertino. It was a four-bedroom ranch style house on Presidio Drive, close to Apple's first offices. Steve told me that he didn't want to get a house with just the two of us because it felt insufficient to him.
>
> I had suggested to Steve that we separate, but he told me that he just couldn't bring himself to say goodbye. I was glad to hear this but I was also, by this time, deferring to his ideas way too often.
>
> —Chrisann Brennan[34]

In the shared house, Jobs initially claimed the front bedroom, then realized Brennan had a better bedroom. So he moved her possessions out.[35] At Apple, Brennan would hold a similar position on the Apple II as Jobs's adopted sister Patricia filled on the original Apple computer.

> At Apple I worked in the shipping department where, if I remember correctly, I soldered disconnected chips onto boards

> and also screwed those same boards into Apple II cases for final assembly.
>
> —Chrisann Brennan[36]

Jobs would later admit to already being a millionaire at the time his first daughter was born. Yet through late 1982, he would repeatedly deny being Lisa's father, in court proceedings and later, in the media. Brennan would allege that Jobs's lawyers even drew blueprints of the shared house in order to suggest that any male could be the father of their child.[37] Suppressing information about his paternity resulted in the release of the information at a moment in time when Jobs was being considered for a major recognition from a magazine, harming his ambitions.

ATTENTION TO DETAIL

Jobs's father Paul had stressed that an individual should give the same level of effort to hidden parts of a project (the back of the fence) as well as the part that others would see. When asked whether customers really cared about the level of care taken by Apple to make the inside of the computer case attractive, Jobs was resolute:

> Woz and I cared from the very beginning. And we felt the people who were going to own the Apple II would care, too. We were selling these things for $1,600, I think, which was a lot of money back in 1977, and these were people who generally didn't have $1,600. I know people who spent their life savings on one. Yeah, they cared what it looked like on the inside.
>
> —Steve Jobs[38]

Chapter 5

NOW TRULY WEALTHY

I was worth about over a million dollars when I was twenty-three and over ten million dollars when I was twenty-four, and over a hundred million dollars when I was twenty-five and … it wasn't that important … because I never did it for the money.

—Steve Jobs[1]

APPLE'S 1980 INITIAL PUBLIC OFFERING (IPO)

Jobs may not have done the work for the money but Apple generated a lot of money. Just three-and-a-half years after introducing the Apple II, the corporation was valued at nearly $1.8 billion.

> The time and money each of the three sacrificed to make Apple a success were amply rewarded on December 12, 1980, when underwriters Morgan Stanley and Hambrecht & Quist took the company public. Originally filed to sell at $14 a share, the stock opened at $22 and all 4.6 million shares sold out in minutes. The stock rose almost 32 percent that day to close at $29, giving the company a market valuation of $1.778 billion. Jobs, the single largest stockholder with 7.5 million shares, suddenly had

> a net worth exceeding $217 million. Not too shabby for a college dropout. Woz's 4 million shares were worth a respectable $116 million. Pretty good for a wire-head who never wanted to build a company. Even Markkula couldn't complain. His 7 million shares were valued at $203 million, for an unbelievable 55,943 percent annualized return on his original 1977 stake!
>
> —*Apple Confidential*[2]

Apple sold shares of stock to the public in December 1980 at $22 per share.[3] An investor who purchased 100 shares in the IPO for $2,200 and never sold them, would have 5,600 shares worth almost $600,000 just 35 years later. Results like this suggest that Apple has been a phenomenal success and created a lot of wealth, which is indeed true. However, the growth and success of Apple was not consistent. In fact, the company almost failed multiple times, including a situation that was so dire Apple would later make an agreement with Microsoft to ensure sufficient funding.

College friend, India companion, and early Apple employee (number 12) Dan Kottke was one of the notable individuals who did not become wealthy from the IPO, although he had tried to speak to Jobs after never receiving a stock option grant despite being with the firm from the beginning. In fact, Kottke later stated that his college friend Jobs simply stopped talking to him—completely—in 1980.

> It got to be the summer of 1980 and I never had a stock option. No one would ever talk to me about it. All I wanted was just to touch base with Steve about it, and he just would not talk to me. He kept me waiting outside his office for hours, on multiple occasions. It was very cold. And you know how he is, he would just be on the phone endlessly until I went away, because he didn't want to talk to me.
>
> —Dan Kottke[4]

WOZ'S PLANE CRASH

Fewer than two months after the IPO, Apple co-founder Steve Wozniak crashed a small plane (Beech A36TC) he was unqualified to operate, injuring himself and passengers on February 7, 1981.[5] For five weeks after the crash, Woz was completely unable to create new memories. This event led to Wozniak's first departure from Apple. Woz left Apple, promoted some concerts, and came back two years later to help the Apple II team.

> After my plane crash I had a 5-week period where I didn't remember from minute to minute and didn't know that time had passed. I had all my old memories and got around but was a little weird to friends and family who didn't know about this forward form of memory loss.
>
> —Steve Wozniak[6]

The absence of Wozniak for those two years allowed focus to flow from the commercially successful Apple II to other projects; in 1985, Woz would claim the company had "been going the wrong direction for the last five years." The plane crash resulted in the person who generated the initial Apple and Apple II products leaving the firm for a few years; he was more than just another employee. At that time, the company was solely reliant on products Wozniak had created for revenue.[7] The Apple II, released in 1977, was still effectively all of Apple's revenue four years later. Jobs was working on other projects that demanded his attention during this period, including LISA and then Macintosh.

In mid-1981, Michael Scott quit Apple and initial investor Mike Markkula stepped in as CEO for the next two years; Jobs had still never been CEO of the firm he co-founded.

IBM INTRODUCES THEIR FIRST PC, RUNNING MICROSOFT DOS

The day the IBM PC was announced was a critical point in the history of Apple, although Microsoft had delivered the operating system for IBM. Microsoft had purchased a version of a Disk Operating System (DOS) that was then rebranded to use on the IBM PC and compatibles. On the day of the IBM announcement in 1981, Bill Gates happened to be at Apple headquarters.[8] Of the personnel at Apple looking at the announcement: "They didn't seem to care," he said. "It took them a year to realize what had happened."[9]

Although Markkula had always assumed Apple would face competition from IBM, the company actually expected to have major competition from IBM in the consumer market sooner. By 1981, Apple was established as the major computer company for the average user in both homes and schools. So Apple and Jobs, in a bit of hubris, welcomed IBM as a personal computer manufacturer in a full-page newspaper advertisement created by agency CHIAT\DAY.

A COMPUTER NAMED LISA

Apple's first attempt at a household successor to the Apple II was aimed at consumers (not businesses) but had a price tag only major corporations could afford—at $10,000, this could exceed an annual salary in some industries. While the project attempted to commercialize some of the ideas Jobs had "found" at Xerox PARC, the most noteworthy part of the product was the name, The Lisa. Isaacson probed Jobs in order to derive the origin of the name.

> The name Jobs chose for it would have caused even the most jaded psychiatrist to do a double take: the Lisa. Other computers had been named after daughters of their designers, but Lisa was a daughter Jobs had abandoned and had not yet fully admitted was his. "Maybe he was doing it out of guilt," said Andrea Cunningham, who worked at Regis McKenna on public relations for the project. "We had to come up with an acronym so that we could claim it was not named after Lisa the child." The one they reverse-engineered was "local integrated systems architecture," and despite being meaningless it became the official explanation for the name. Among the engineers it was referred to as "Lisa: invented stupid acronym." Years later, when I asked about the name, Jobs admitted simply, "Obviously it was named for my daughter."[10]

Lisa? At the time, observers of Apple products knew of The Lisa (a soon-to-be-released and poorly-selling computer). Observers did not know about Lisa Brennan (Steve Jobs's child), who was already four years old when promotional materials for the computer were released.

And Lisa, like a previous Apple III intended for business, failed in reaching their intended audiences. Apple developed the reputation of being designed for individuals and schools, a reputation that wasn't fully taken advantage of until 15 years later.

1982 *TIME* INTERVIEW

Jobs remained the public face of the firm, although the CEO was still Mike Markkula. He had multiple high-profile interviews in the early 1980s while at Apple; he was even a candidate for *Time* magazine's "Man of the Year" honor. *Time* sent Michael Moritz to complete an interview of Jobs, with Jay Cocks writing the profile of Jobs.

In the *Time* article published the first week of 1983, Guy Tribble introduces the concept of Jobs's "reality-distortion field" to the general public, suggesting that Jobs:

> ... has the ability to make people around him believe in his perception of reality through a combination of very fast comeback, catch phrases and the occasional very original insight, which he throws in to keep you off balance.
>
> —Guy Tribble[11]

Jobs had temporarily dropped vegetarianism at the time, saying "Interacting with people has got to be seriously balanced against living a little healthier ... The amount of time you spend shopping and preparing and eating food is enormous ... The amount of energy your body spends digesting the food in many cases exceeds the energy we get from the food." He also discussed using LSD in the popular media, noting of his experience "All of a sudden the wheatfield was playing Bach."[12]

Then came the bombshell, and the denial. *Time* revealed that Jobs was the father of a daughter named Lisa but that Jobs had persistently denied paternity of his daughter. *Time* also picked up that Jobs, curiously, had recently named his most recent project, which was to be released the next month, "The Lisa." Of Lisa Brennan, the article noted:

> The baby, a girl, was born in the summer of 1978, with Jobs denying his fatherhood and refusing to pay child support. A voluntary blood

> test performed the following year said "the probability of paternity for Jobs, Steven… is 94.1%." Jobs insists that "28% of the male population of the United States could be the father." Nonetheless, the court ordered Jobs to begin paying $385 a month for child support. It may be noted that the baby girl and the machine on which Apple has placed so much hope for the future share the same name: Lisa.[13]

The *Time* article also mentions Jobs's autocratic style, ignoring the needs and desires of others with a manner that did not accept questioning or dissent.

> He would have made an excellent King of France.
>
> —Jef Raskin[14]

THE MAN OF THE YEAR WAS NOT A MAN (OR WOMAN, OR PERSON, OR GROUP)

Jobs subjected himself to the *Time* interview because he believed he was going to be named the magazine's "Man of the Year." *Time* then did something that had not been done in the previous 55 years of naming a "Man of the Year." Instead of naming a person or group of people, *Time* decided to name "The Computer" as "Device of the Year" on January 3, 1983, in the same issue as the published Steve Jobs interview. Jobs was irate with the direction the article had gone, including the first major public release that he had a daughter out of wedlock. As the interviewer would later declare,

> Steve rightly took umbrage over his portrayal and what he saw as a grotesque betrayal of confidences, while I was equally distraught by the way in which material I had arduously gathered for a book about Apple was siphoned, filtered, and poisoned with a gossipy benzene by an editor in New York whose regular task was to chronicle the wayward world of rock-and-roll music. Steve made no secret of his anger and left a torrent of messages on the answering machine. … He, understandably, banished me from Apple and forbade anyone in his orbit to talk to me. The experience made me decide that I would never again work anywhere I could

> not exert a large amount of control over my own destiny or where I would be paid by the word.
>
> —Michael Moritz[15]

Moritz had done the reporting but not the editing, although Jobs felt Moritz was responsible. Moritz was in the process of writing an early book about Apple; Jobs restricting him from the Apple campus, and not permitting his circle of friends to talk with Moritz, certainly made the process of writing Moritz's book more difficult. But through Jobs's anger, Moritz decided he should never undertake work where he didn't control his own destiny. And, like others who were within Jobs's orbit, the author Moritz later also becomes a billionaire himself, as a venture capitalist like Arthur Rock.

The bad experience with *Time* did not prevent him from undertaking additional interviews, as he remained the engaging public face of Apple. Jobs even sat for later *Time* interviews, albeit more than a decade later. The next major profile would be released in 1985, after the introduction of Macintosh and while fighting for his position within the company he had co-founded.

JUNE 1983

Apple's partner Microsoft was still creating software for Apple, and realized that the common operating systems of the day, which were text-based and lacking mice, were no longer worth investment at all. In what is called the Applications Strategy Memo, Bill Gates and Steve Ballmer announced:

> Microsoft believes in mouse and graphics as invaluable to the man-machine interface. We will bet on that belief by focusing new development on the two new environments with mouse and graphics, Macintosh and Windows…
>
> Microsoft will not invest significant development resources in new Apple II, MSX, CP/M-80 or character-based IBM PC applications. We will finish development and do a few enhancements to existing products.
>
> —Bill Gates and Steve Ballmer[16]

As a very integrated partner with Apple, this meant Microsoft would be releasing software for Macintosh before any versions of Microsoft Windows were ever released. Bill Gates even proffered that Apple was very likely right in the initial decision to abandon the existing models of operating system for something most computer users would recognize today as part of Macintosh.

> The Mac was a very, very important milestone. Not only because it established Apple as a key player in helping to find new ideas in the personal computer but also because it ushered in a graphical interface. People didn't believe in graphical interface. And Apple bet their company on it, and that is why we got involved in building applications for the Macintosh early on. We thought they were right.
>
> —Bill Gates[17]

APPLE WINS THE FIRST INTELLECTUAL PROPERTY LAWSUIT

On August 30, 1983, Apple won the first major intellectual property lawsuit. In *Apple Computer, Inc. v. Franklin Computer Corporation*, Franklin admitted that it had copied software written by and for Apple, largely because it would have been too time-consuming and expensive to write the software on their own:

> Franklin did not dispute that it copied the Apple programs. Its witness admitted copying each of the works in suit from the Apple programs. Its factual defense was directed to its contention that it was not feasible for Franklin to write its own operating system programs.
>
> —U.S. Court of Appeals[18]

From that date forward, there was no doubt that copyrighted software was fully protected under the laws of the United States. For Apple, this development should have been a cause for celebration—the company had the single most widely-sold computer in the world at the time (Apple II series) and was to release Macintosh shortly, with the ability to vigorously defend their software and innovations from copyright infringement by competitors.

Chapter 6

OUSTER FROM APPLE

How can you get fired from a company you started? Well, as Apple grew we hired someone who I thought was very talented to run the company with me, and for the first year or so things went well. But then our visions of the future began to diverge and eventually we had a falling out. When we did, our Board of Directors sided with him. So at 30 I was out. And very publicly out. What had been the focus of my entire adult life was gone, and it was devastating.

—Steve Jobs, at Stanford's 2005 Commencement[1]

HIRING OF JOHN SCULLEY IN 1983

There's an expression that hindsight is 20/20—one can see clearly what should have been done in the past once the future implications of an action are known. In 1996, Jobs was able to admit that he made a massive error in 1983.

Steve Jobs (in 1996): Ehm what can I say? I hired the wrong guy.
Q: That was Sculley?
Steve Jobs (in 1996): Yeah and eh he destroyed everything I spent ten years working for. Ehm starting with me but that wasn't the saddest part. I would have gladly left Apple if Apple would have turned out like I wanted it to.[2]

With more than a decade of reflection, and then only involved with future creations NeXT and Pixar, Jobs came up with "I hired the wrong guy." But in 1983, Sculley was seen as one of the best remaining executives to lead an Apple that was growing rapidly into a major manufacturing business. Sculley was leading Pepsi, which had a reputation as an upstart against an established firm; Sculley told Jobs that research had shown that families were embarrassed to serve Pepsi to guests, bringing cans or bottles of Coca-Cola from the kitchen to the table but pouring Pepsi into glasses before serving to guests. And Jobs, always excellent at selling a vision, famously asked Sculley if he wanted to sell sugar water (at PepsiCo) or if he wanted to make a real difference in the world by joining Apple.

> And then he looked up at me and just stared at me with the stare that only Steve Jobs has and he said do you want to sell sugar water for the rest of your life or do you want to come with me and change the world and I just gulped because I knew I would wonder for the rest of my life what I would have missed.
>
> —John Sculley[3]

Apple had a new CEO who would remain with the firm for a decade, with varying levels of success. Initially Sculley and Jobs were reasonably good friends. Jobs would not become CEO of Apple until much later. Three decades later, Sculley would admit that movies released about Steve Jobs didn't truly reflect his personality, even his (occasional) genuine friendliness. Jobs cared about the people he worked with but only wanted the most exceptional individuals within his sphere.

> Part of his personality was he was a passionate perfectionist, but there were so many other parts of Steve's personality that I knew because Steve and I were not only business partners, but we were incredibly close friends for several years. … I could tell you that the young Steve Jobs that I knew had a great sense of humor. He was on many occasions, when we were together, very warm. He cared a lot about the people he worked with and he was a good person.
>
> —John Sculley[4]

APPLE'S EVENTUAL FUTURE IS MACINTOSH

At the end of 1983, Jobs is preparing to release Macintosh, a product that he's sure will be a hit and revolutionize computing. However, Jobs stepped on toes on the way to the release. He moved to Macintosh after being removed from The Lisa, to a project that was already well-formed by an individual named Jef Raskin.

> Then there was the whole episode of the Macintosh. Jef Raskin was ... bitter about the way Steve treated him until the end of his life. Jef was a renaissance man in so many areas. Steve has a debt of gratitude to Jef in a personal growth kind of way. And it was Jef who clearly got that Mac project started.
>
> —Dan Kottke[5]

With Raskin no longer at Apple, Jobs worked with Bill Gates and Microsoft to develop the operating system and applications for the launch of Macintosh. In fact, Microsoft had more individuals working on Macintosh than Apple had working on Macintosh.

> It's a great machine ... It allows us to write software which is significantly easy to use ... There's no way this group could have done any of this stuff without Jobs.
>
> —Bill Gates

Reflecting later, Gates recognized again Jobs's contributions to the product beyond those initiated by Jef Raskin. Gates also suggested that working with Jobs was an enjoyable experience.

> We worked closely with Apple throughout the development of the Macintosh. Steve Jobs led the Macintosh team. Working with him was really fun. Steve has an amazing intuition for engineering and design as well as an ability to motivate people that is world class.
>
> It took a lot of imagination to develop graphical computer programs. What should one look like? How should it behave? Some ideas were inherited from the work done at Xerox and some were original.
>
> —Bill Gates[6]

While the Macintosh was under development, Jobs was maintaining a relationship with Joan Baez, a popular folk singer who was 14 years older than he.

JOAN BAEZ

We met in Palo Alto (California)... and we were close in the late '70s and early '80s. After that we always kept in touch, we'd call or email. I'd always snag a new laptop out of him!

—Joan Baez[7]

Although Jobs tended to maintain absolute secrecy on his projects, he brought Baez and her sister to see Macintosh while the product was under development.

> Not only did he tell them about our Macintosh-in-development but he decided to SHOW it to them too. We sat there doubly dumfounded at the disclosure of our secret project to an outsider ... who happened to be a huge celebrity ... that we actually got to meet! Hopefully Steve had them sign a non-disclosure agreement, but I never saw it.
>
> —Jerry Manock[8]

When Baez released her memoirs in 1987, she mentioned her former romantic interest in the acknowledgments and that he had given her a computer; she did not reveal which Apple product she received from Jobs:

> Steve Jobs for forcing me to use a word processor by putting one in my kitchen.
>
> —Joan Baez[9]

1984 (THE YEAR OF MACINTOSH)

In early January 1984, Michael Moritz's book about Apple was released, although Jobs was still incensed about the *Time* incident.

> Steve and I had a somewhat tortured relationship. I wrote the first book about Apple and it came out in about 1984 just before the

> Macintosh computer. I got to know him extremely well— I knew him probably back then as well as anyone outside of his immediate family.
>
> —Michael Moritz[10]

During the preparation for Macintosh production, Jobs the artist had an idea that hearkened back to his father telling him to make the unseen side of the fence as attractive as the visible side and the episode of designing the interior of the Apple II to be visually pleasing. Those with original Macintosh computers had the names of the design team engraved on the inside of the computer case, as Macintosh was considered a work of art by Jobs.

> The Mac team had a complicated set of motivations, but the most unique ingredient was a strong dose of artistic values ... First and foremost, Steve Jobs thought of himself as an artist, and he encouraged the design team to think of ourselves that way, too ... Since the Macintosh team were artists, it was only appropriate that we sign our work. Steve came up with the awesome idea of having each team member's signature engraved on the hard tool that molded the plastic case, so our signatures would appear inside the case of every Mac that rolled off the production line.
>
> —Andy Hertzfeld[11]

1984 (THE ADVERTISEMENT)

It is now 1984. It appears IBM wants it all. Apple is perceived to be the only hope to offer IBM a run for its money. Dealers initially welcoming IBM with open arms now fear an IBM dominated and controlled future. They are increasingly turning back to Apple as the only force that can ensure their future freedom. IBM wants it all and is aiming its guns on its last obstacle to industry control: Apple. Will Big Blue dominate the entire computer industry? The entire information age? Was George Orwell right about 1984?

—Steve Jobs[12]

The "1984" advertisement is regarded as one of the most iconic television advertisements of all time, depicting hundreds of individuals told how to think until an athletic female destroys—with a sledgehammer—the screen of the individual giving instructions. The advertisement was to introduce Macintosh, which was promoted as changing how individuals would use computers forever. The advertisement itself had a short history and was only shown twice, December 1983 and January 1984.

Steve Hayden of CHIAT\DAY authorized the production of the advertisement and years later, wrote about the untold story behind it. The short version of the advertisement had been played on an Idaho television station in 1983 in order to make the commercial eligible for awards given to advertisements in 1984. The short version was also run "in the top 10 U.S. markets, plus, in an admittedly childish move, in an 11th market—Boca Raton, Fla., headquarters for IBM's PC division." Jobs spearheaded the advertisement but it almost didn't run at all, as Hayden reflected upon later:

> The spot had a brush with death after Mike Murray [member of Macintosh team] and Jobs played the spot for the Apple board of directors in the fall of 1983. When the lights came up, Murray reported that most of the board members were holding their heads in their hands, shaking them ruefully. Finally, the chairman, Mike Markula [sic], said, "Can I get a motion to fire the ad agency?"
>
> They absolutely hated the spot to a man, and they were all men in those days, but left it to Jobs and Sculley whether to run it or not. Apple co-founder Steve Wozniak saw the spot and offered to pay half the cost of running it out of his personal checking account.[13]

The advertising agency was told to sell off the commercial time that was reserved for Super Bowl XVIII; the group sold two of the small ad slots but left the longer slot for the full version of the ad. Even then, Steve Jobs still had doubts about using the Super Bowl as a venue to showcase what would become their most famous advertisement. Steve Jobs had a rather peculiar concern about whether Apple should be advertising during the Super Bowl.

> I don't know a single person who watches the Super Bowl.
>
> —Steve Jobs[14]

INTRODUCING MACINTOSH

An eloquent speaker, Jobs introduced Macintosh to the world as the third major product in the history of personal computing.

> There have only been two milestone products in our industry—the Apple II in 1977 and the IBM PC in 1981. Today ... one year after LISA we are introducing the third industry milestone product. ... Macintosh. Many of us have been working on Macintosh for over two years now and it has turned out insanely great. You've just seen some pictures of Macintosh now I'd like to show you Macintosh in person.
>
> —Steve Jobs[15]

In terms of communications, Jobs spoke about how innovations rapidly changed in the late 1800s. With telegraphs, sending a message across the country required specialists and the messages still took 3–4 hours to repeat across the country. Learning Morse Code to even begin use the telegraph was simply too complicated, so everyone had to rely upon telegraph operators. With telephones, potential customers already knew how to use it, although the cost meant that for the first 10 years, about 200,000 people had access to phones. And telephones with voice could relay meaning better than any telegraph ever could. He thought the existing computers of the day, including IBM-compatibles and even the Apple IIs, were effectively the telegraph. Macintosh was the intuitive telephone that enhanced communication for everyone.

> And what we think we have here is the first telephone. And in addition to letting you do the old spreadsheets and word processing, it lets you sing. It lets you make pictures. It lets you make diagrams where you cut them and past them into your documents. It lets you put that sentence in Bold Helvetica or Old English, if that's the way you want to express yourself.
>
> —Steve Jobs[16]

Macintosh, with the support of Microsoft, was released in January 1984 as the second Apple product that would revolutionize computing. Jobs was right about the "insanely great" Macintosh; he would just run out of time at Apple before the product sold as if it truly were "insanely great." Those who did interact with Macintosh early realized that the computer was different and revolutionary.

> Finally, the day came that changed my life forever. I went to a computer store in downtown Washington, D.C., to see my first Mac in person. At first the shop appeared deserted, until I rounded a corner and there were about 30 people in a cluster, of course huddled around the only Mac on the floor. I waited patiently for my turn. When I sat down, I drew a tree in Mac Paint and printed it out on a companion Apple ImageWriter printer. This was the first time that what you drew on the screen would actually print out and look exactly the same. This was big. All of a sudden, I just knew this was important. This was going to change the world. I wanted to meet all the people whose signatures were molded in plastic on the inside of the Macintosh: the actual creators, including Steve. At that point, I became obsessed.
>
> —Tom Reilly[17]

1984 *NEWSWEEK* INTERVIEW

Jobs sat for an interview with *Newsweek* after the introduction of Macintosh in 1984. He provided general musings as well as responses to detailed questions about his firm. He reveals that he was ambidextrous after noting that most of technical people he knew were left-handed. On working in the industry:

> A lot of people will probably take this analogy wrong, but there are a good number of people who would have loved to have had even the most menial job on the Manhattan Project in Los Alamos and watch those brilliant people work together for that period of time.
>
> —Steve Jobs[18]

Following that up, he referred to making fantastic products as more important than concepts such as work-life balance; fantastic products can only be made at set moments in time, and those overrode concerns like being able to use one's own kitchen.

> There are moments in history which are significant and to be a part of those moments is an incredible experience. In other

words, there are more important things than cooking in your own kitchen.

—Steve Jobs[19]

He acknowledged the extensive demands of working at Apple, on the employees as well as families. Jobs felt the compensation was financially rewarding but that it did require some of the staff to periodically sell shares for major purchases (like homes) to show their families that the sacrifice was worthwhile.

One of the trends I've seen is that once things seem a little stable, once the company has made it over some critical hurdles, some of the people will sell enough of their stock to buy a house or do something which may not mean that much to them, but will mean something, let's say, to their spouse or to their family, which hasn't seen enough of them for the last two years. They'll want to do something to sort of say, "Hey, you know, what I've been working on really has been valuable, it really has been worth it and besides my loving it, it has produced something for the family or for both of us."

—Steve Jobs[20]

1985 *PLAYBOY* INTERVIEW

Jobs was interviewed by the adult magazine *Playboy* in 1985, where he really articulated what the computers of 1985 (such as Macintosh) could do and what computers of the future would be able to do.

It's very crude today, yet our Macintosh computer takes less power than a 100-watt light bulb to run and it can save you hours a day. What will it be able to do ten or 20 years from now, or 50 years from now?

—Steve Jobs [21]

From the initial point of time savings, he saw computers as being able to automate tasks for average citizens, which was revolutionary at the time.

We're going to be able to ask our computers to monitor things for us, and when certain conditions happen, are triggered, the computers will take certain actions and inform us after the fact ... Simple things like monitoring your stocks every hour or every day. When a stock gets beyond set limits, the computer will call my broker and electronically sell it and then let me know. Another example is that at the end of the month, the computer will go into the data base and find all the salesmen who exceeded their sales quotas by more than 20 percent and write them a personalized letter from me and send it over the electronic mail system to them, and give me a report on who it sent the letters to each month. There will be a time when our computers have maybe 100 or so of those tasks; they're going to be much more like an agent for us. You're going to see that start to happen a little bit in the next 12 months, but really, it's about three years away. That's the next breakthrough.

—Steve Jobs[22]

And Jobs spoke about how Apple reached out to colleges—including every Ivy League school—to get each of those schools to purchase at least 1,000 Macintosh computers, automatically making Macintosh the default computer in higher education (as the Apple II series still was in K–12 education).

We also wanted Macintosh to become the computer of choice in colleges, just as the Apple II is for grade and high schools. So we looked for six universities that were out to make large-scale commitments to personal computers—by large, meaning more than 1,000 apiece—and instead of six, we found 24. We asked the colleges if they would invest at least $2,000,000 each to be part of the Macintosh program. All 24—including the entire Ivy League—did. So in less than a year, Macintosh has become the standard in college computing. I could ship every Macintosh we make this year just to those 24 colleges. We can't, of course, but the demand is there.

—Steve Jobs[23]

Steve Jobs also agreed with the interviewer that the folks making "insanely great" computers might be a little bit insane themselves.

And he insisted that no other billion-dollar computer companies would ever progress through a garage again, with Apple as—in his mind—the only feasible source for any future innovations in computing. He learned about one use of the Apple II that displeased him, although he was pleased that if it had to be any personal computer, he was happier to know that the computer used was an Apple II rather than a competitor.

> I saw a videotape that we weren't supposed to see. It was prepared for the Joint Chiefs of Staff. By watching the tape, we discovered that, at least as of a few years ago, every tactical nuclear weapon in Europe manned by U.S. personnel was targeted by an Apple II computer. Now, we didn't sell computers to the military; they went out and bought them at a dealer, I guess. But it didn't make us feel good to know that our computers were being used to target nuclear weapons in Europe. The only bright side of it was that at least they weren't [Tandy–Radio Shack] TRS-80s! Thank God for that.
>
> —Steve Jobs [24]

WHERE WAS APPLE AT THE TIME?

In 1985, the company was split between a number of pricing and marketing decisions. The Apple II series, originally released in 1977, was still effectively all of Apple's revenue, despite effort on Apple III, LISA, and Macintosh. As a result, the company had to spend advertising dollars and effort on the Apple II, while Jobs insisted the company spend money advertising the Macintosh and sell the Macintosh for $500 less than the price selected by Sculley and Apple's board of directors. There was conflict between Jobs and Sculley, with low probability that both parties would get what each desired.

And the Macintosh that hadn't taken off yet? This disconnect led to Jobs's departure from the firm. Sculley later admitted that upon his departure from Apple eight years later in 1993, the Macintosh had indeed become effectively all the revenue of the firm. The first and second major product lines, Apple II series and Macintosh, were Apple's primary revenue drivers for more than two decades.

FIRST RESIGNATION FROM APPLE

In his resignation letter from Apple of September 17, 1985, Jobs discussed starting a new venture, the company that would become NeXT. He presented specific concerns about his role with Apple at the time; even as Chairman of the Board at the time, he was not even allowed to see basic management reports:

> As you know, the company's recent reorganization left me with no work to do and no access even to regular management reports.
>
> —Steve Jobs[25]

And with that, Jobs was gone from Apple. He did meet with some trusted Apple employees before leaving. While wearing blue jeans and no shoes, Jobs is said to have remarked:

> I don't wear the right kind of pants to run this company.
>
> —Steve Jobs[26]

Apple was less than a decade old, but employees of the firm recognized already the contributions Jobs had made to the firm. One employee even compared the departure of Steve Jobs from Apple as similar to November 22, 1963, the day President John F. Kennedy was assassinated in Dallas, Texas.

> I remember distinctly the day of that board meeting where he was ousted, and it had the quality of the day Kennedy was shot. Without him, it just wasn't the same company. When you look back, it was something of a miracle the company survived until he returned.
>
> —Former Apple Manager[27]

In the immediate timeframe afterward, Jobs's perception took a hit. Writers made suggestions that the true entrepreneurs weren't necessarily the ones who were able to lead companies to be their largest and most successful. As Tushman wrote, "the very characteristics that lead entrepreneurs to start companies—independence, innovation and a commitment to ideas—are the same ones that can cause their demise as managers. A mature firm cannot tolerate relentless turmoil or a tendency to dash off in all directions."[28]

Jobs and Wozniak felt they had to start the Apple product, and then the company, precisely when they did. Looking back, Jobs knew there was little possibility of starting a year earlier or later with the Apple.

> Apple was unique ... if we had tried it one year earlier or one year later, it wouldn't have happened.
>
> —Steve Jobs[29]

How did Apple stock respond to Jobs's resignation from the company he co-founded? Upon his departure, the stock gained sharply in value. Jobs removing himself from the firm was viewed positively by the market, and some analysts were on the record as saying Jobs's departure was actually a good development for the firm. But Nolan Bushnell of Atari, who hired Jobs in the early 1970s, immediately questioned who would lead Apple's innovations without Jobs present:

> Says David Gold, a Palo Alto venture capitalist: "It's good news for Apple that he's out of their hair. The loss of a few employees is probably a small price to pay to have Steve Jobs going off and doing something else." But Nolan Bushnell, who founded Atari and subsequently launched and left several other firms, including Pizza Time Theater, is not so sure. Says he: "Where is Apple's inspiration going to come from? Is Apple going to have all the romance of a new brand of Pepsi?"[30]

Good friend Wozniak, who had left in February with much of the rest of the Apple II team, noted:

> Steve can be an insulting and hurtful guy.
>
> —Steve Wozniak[31]

Sculley admitted that he never patched any form of friendship with Steve Jobs after that point. More notably, he admitted that his forcing Jobs out of the firm completely was a mistake.

> No. Never repaired. And it's really a shame because if I look back I see what a big mistake on my part. See, I came from corporate

> America. There it was kind of secular, there wasn't the passion that entrepreneurs have. I have so much respect now decades later for founders, for the belief and passion and vision that they have. So to remove a founder, even if he wasn't fired, was a terrible mistake. I wish that Steve and I had gotten together again and found even some part of our friendship, but it didn't happen.
>
> —John Sculley[32]

BEFORE THE CALENDAR GOES TO 1986 ...

Jobs wasn't done with 1985, though. He still had to start his next company, which was named ... NeXT. Despite starting a new firm, he had time in his schedule to conduct a major post-Apple interview for *Newsweek*. Although he had been Chairman of the Board and a co-founder of the company, he expressed that he had spoken to Sculley a grand total of three times in the four months before his departure:

> Well, given the fact that I've spoken to him only three times since (May)—that says something about the degree of communication we've had—I don't know what will happen with my relationship with John.
>
> —Steve Jobs[33]

He noted that Sculley, as CEO, had the right to say there wasn't really a role for Jobs as an everyday employee at Apple. He noted that Apple was a $1.5 billion company and was probably destined to be no more than a $5–$10 billion company, where he admittedly would not be the best to lead the company.

He even talked about offers he had received to be a college professor, when he had only taken one semester of college courses for credit.

> I got three offers to be a professor during this summer, and I told all of the universities that I thought I would be an awful professor. What I'm best at doing is finding a group of talented people and making things with them. I respect the direction that Apple is going in. But for me personally, you know, I want to make things. And if there's no place for me to make things there, then I'll do

> what I did twice before. I'll make my own place. You know, I did it in the garage when Apple started, and I did it in the metaphorical garage when Mac started.
>
> —Steve Jobs[34]

When talking about his new initiative (NeXT), he described the idea as potentially being a bit crazy. However, Jobs had a little bit of flexibility in choosing his future activities (even his crazy ones), as his mid-1980s net worth was already well over $100 million.

> We talked about this enterprise, you know, for the first time less than two weeks before I told the board that I wanted to start this company. And we have no business plan. We haven't done anything. Now, you might say we're all crazy. We have a general direction. We want to find out what higher education needs. We plan to go visit a lot of colleges in October and just listen.
>
> —Steve Jobs[35]

Jobs also took the available opportunity to taunt to his former firm, while expressing that his new team would not be taking proprietary information from Apple (although Apple under Sculley's direction sued him anyway).

> There is nothing, by the way, that says Apple can't compete with us if they think what we're doing is such a great idea. It is hard to think that a $2 billion company with 4,300-plus people couldn't compete with six people in blue jeans.
>
> —Steve Jobs[36]

PART 3

LIFE AFTER APPLE (OR THE YEARS BETWEEN)

One of Steve Jobs's first actions after leaving Apple? Using an iconic advertisement created by CHIAT\DAY (welcoming IBM to the personal computing market) as a model to welcome CHIAT\DAY to their own post-Apple timeframe; CHIAT\DAY was removed shortly after Steve Jobs. And as only Jobs could say,

> I'm expecting some new, "insanely great" advertising from you soon.
>
> —Steve Jobs

And there was a life after Apple for CHIAT\DAY. That life after Apple would even include a second life with Apple …

For Jobs?

Robert Frost wrote the poem "The Road Not Taken," which started:

> Two roads diverged in a yellow wood,
>   And sorry I could not travel both …

Jobs found himself with two roads that diverged, neither the first nor second ended in regret, as he skipped the rest of the poem.

> One road was NeXT, which (unknown to him) would lead back to Apple.
>
> The other road was Pixar, which (unknown to him) would lead to Disney.

He took both, and that has made all the difference.

Chapter 7

NEXT WAS NEXT

The NeXT workstation was designed for education, specifically activities that would require a lot of processing power then unavailable on desktops, or to replicate very specialized laboratories and equipment that would otherwise be very costly. Jobs spoke about how he was inspired to take on the new initiative, before he had even left Apple.

> I have a friend at Stanford, a Nobel Prize-winning molecular biologist. He was showing me what some of his students were doing to understand how proteins fold. He asked, "Couldn't you model this stuff on a computer if you had something more powerful than a PC?" It really got me thinking. What if you came up with something that was as easy to use as a Mac, or even easier, and had the power of a workstation? What if you unleashed that machine in higher education? The more I thought about it, the more excited I got.
>
> —Steve Jobs[1]

Early in the development of NeXT, he spoke to educators about the types of activities an advanced computer might permit in the classroom.

> You'd offer a physics student a personal linear accelerator or a ride on a train going the speed of light," he told a group of educators in 1986. "You'd take a biochemistry student and let him experiment in a $5 million DNA wet lab. You'd send a student of 17th-century history back to the time of Louis XIV. Next year we will introduce a breakthrough computer ten to 20 times more powerful than what we have today."
>
> —Steve Jobs[2]

One year after the founding, NeXT was valued as a company technically worth $120 million when Ross Perot—founder of EDS, Perot Systems, 1992 presidential candidate, and one of the entrepreneurs Jobs admired—invested $20 million for 16 percent of the firm.[3] Perot was afraid of missing out on Jobs's second successful venture.

The NeXT workstation, once completed, was an exceptionally artful piece of computer hardware, appreciated even by those with products designed for competitors. "Microsoft Chairman Bill Gates has described the NeXT machine as the most beautiful computer ever built."[4] Given the lack of processing power in the mid-1980s and Moore's Law (which has historically suggested that microprocessors consistently double in power every few years), even Jobs was amazed at how this computer designed for higher education turned out.

> Right. I mean, we had the idea of doing a machine for higher education in the fall of 1985, but our original concept was about a third as good as the computer turned out to be. The improvement came from a lot of interaction between people in higher education and those of us at NeXT.
>
> —Steve Jobs[5]

Jobs spoke eloquently of the NeXT, and described the three major innovations over other computers of the time. The NeXT was powerful and the costs were approximately the same as other personal computers of the days. The NeXT was designed to be connected to a network. And the NeXT was designed with the intent that new applications could be built very quickly.

> It's not so much different than everything I've ever done in my life with computers starting with the Apple II and the Macintosh, and now NeXT, which is if you believe that these are the most incredible tools we've ever built, which I do, then the more powerful tool we can give to people, the more they can do with it. And in this case we found a way to do two or three things that were real breakthroughs. Number one was to put a much more powerful computer in front of people for about the same price as a PC. The second was to integrate that networking into the computer so we can begin to make this next revolution with interpersonal computing. And the PCs so far have not been able to do that very well. And the third thing, and maybe the most important, was to create a whole new software architecture from the ground up, that lets us build these new types of applications and let's us build them in 25 percent of the time that it normally takes to do on a PC.
>
> —Steve Jobs[6]

While Jobs was considered to be more "mellow" at Pixar, he was still ranked one of America's "Toughest Bosses" by *Fortune* magazine. He required his engineers to build an expensive magnesium casing for the NeXT, then insisted the cost should really be one-tenth of what NeXT ended up paying.

> A manufacturing manager started explaining that making its black, space-age magnesium shell would cost $200 a unit or more. After roughly two minutes, Jobs, his face turning red, cut the startled man off in mid-sentence and began screaming wildly that the shell had to cost $20, that the manager didn't know what he was talking about, and that he was going to ruin the company. The expletive-laden tirade lasted three to four minutes. Of course, the shell of Next's ill-fated computer ended up costing $200. Says a former employee who was present at the meeting: "Tell Steve you

> can't do something because it violates the laws of physics, and he says that's not good enough."[7]

Jobs was also seen as continuing his insistence of getting the answer he wanted, even if he had to change the questions along the way.

> Steve micromanages minute details. He gets a picture in his mind's eye and focuses on it. It's hard to tell him what he doesn't want to hear. He'll keep changing the question until he gets the answer he wants.
>
> —Dan'l Lewin[8]

As had occurred with the Apple II case design, Jobs's obsessive nature about the smallest possible detail arose again. An employee had to go through 37 different shades of green in order to get Jobs's approval on the coloring of a single part of the logo.

> What's especially unnerving, some employees say, is that Jobs sometimes can't— or won't—explain what his expectations are. For example, before the introduction of the NeXT Computer at San Francisco's Davies Symphony Hall in 1988, Jobs made a worker go through 37 different shades of green until she found the one that was right for the company's corporate color. This person describes the final choice as "Steve green." Says the employee: "It was exasperating. The hardest thing was that I couldn't guess what was in his head. I wanted to say, "Oh, come on. Green is green!"[9]

While Jobs was worried about the shade of green, he had met someone more extreme in the vision of correct while having the logo for NeXT designed. Of the designer Paul Rand, who had also designed the logos of companies such as ABC, IBM, and UPS:

> He personally works on perfecting the exterior of a curmudgeon. I think he's perfected it to new heights actually. It's his way of dealing with the part of the world he doesn't necessarily want to deal with … [For the design of the NeXT logo] I asked him if he would

> come up with a few options. And he said, No. I will solve your problem for you. And you will pay me … If you want options, go talk to other people. But I'll solve your problem for you the best way I know how. And you use it or not.
>
> —Steve Jobs[10]

1989—THE ENTREPRENEUR OF THE DECADE

Jobs was named Entrepreneur of the Decade by *Inc.* magazine. The rationale for the decision was not very detailed, and appeared to be due to his popularity rather than entrepreneurial traits. Not only did *Inc.* say they knew little about Jobs before the interview, they also told Jobs that he seemed more interested in business than they had expected.

> *Interviewer:* This is just a personal observation. You seem much more interested in business than we had expected.
> *Steve Jobs:* Business is what I do.[11]

CONTINUED STRUGGLES FOR NEXT

Applications for the NeXT could be built very quickly, but selling the most beautiful computer ever built doesn't help if the software being developed is still insufficient. Five years after leaving Apple, there was a concern in 1990 that much of the promised functionality of NeXT would be vaporware—software that was talked about but did not exist (often to mislead competitors). And this beautiful machine designed by Jobs was soon to fail, as "without many programs to run, the black cube hasn't sold well."[12]

A year later, the company was still struggling. The company had to reduce its workforce and was counting on increased sales to eventually lead to profit. A *Newsweek* article noted the dire circumstances.

> Times are dicey for the industry—and analysts wonder if higher sales will translate into profits at NEXT, which recently laid off 5 percent of its 600 workers. But Jobs could be delivering on his early promises. He predicts a respectable $140 million in sales of 1991 models. He intends to take the company public within the next 18 months, which has some analysts drooling.[13]

Even after six years of challenges and a lack of profitability, investors and analysts still wanted to own part of the company. NeXT not finding immediate commercial success was a stark contrast to Apple, both the initial Apple partnership and then the corporation. The Apple partnership with Wozniak (and briefly Ron Wayne) was profitable in the single year (1976); the Apple corporation was profitable in the first year (1977). Steve Jobs's vision of NeXT was appearing to fail after more than half a decade.

Jobs failed at building a great computer that would take the world by storm. And despite a lack of software applications for NeXT, the software built—including the operating system—was revolutionary.

THE *ROLLING STONE* INTERVIEW

Nine years after leaving Apple, Jobs declined to speak about his former company in a major *Rolling Stone* interview, yet he was willing to say that Apple's inaction was the reason Microsoft had caught up to Apple (and hence understood why Apple was struggling).

> They [Microsoft] were able to copy the Mac because the Mac was frozen in time. The Mac didn't change much for the last 10 years. It changed maybe 10 percent. It was a sitting duck. It's amazing that it took Microsoft 10 years to copy something that was a sitting duck. Apple, unfortunately, doesn't deserve too much sympathy. They invested hundreds and hundreds of millions of dollars into R&D, but very little came out. They produced almost no new innovation since the original Mac itself.
>
> —Steve Jobs[14]

> *Interviewer:* You mentioned the Apple earlier. When you look at the company you founded now, what do you think?
> *Steve Jobs:* I don't want to talk about Apple.[15]

He was willing to talk about Bill Gates, both in terms of his friendship and his belief that Gates had directly stolen ideas from his team.

> The goal is not to be the richest man in the cemetery. It's not my goal anyway.
>
> —Steve Jobs[16]

> I think Bill Gates is a good guy. We're not best friends, but we talk maybe once a month ... I think Bill and I have very different value systems. I like Bill very much, and I certainly admire his accomplishments, but the companies we build were very different from each other.
>
> —Steve Jobs[17]

And some observers began to believe that Jobs was finally receiving his comeuppance as a failure, a form of karma for the Zen Buddhist:

> Remember, this is a guy who never believed any of the rules applied to him ... Now, I think he's finally realized he's mortal, just like the rest of us.
>
> —A former colleague[18]

NEXT COMPUTERS ARE POPULAR ... ON THE SECONDARY MARKET

Jobs spoke about the NeXT computer that was no longer produced, after the company moved solely into software:

> The machine was the best machine in the world. Believe it or not, they're selling on the used market, in some cases, for more than we sold them for originally. They're hard to find even today. We haven't even made them for two, two and a half years.
>
> —Steve Jobs[19]

With NeXT no longer selling computers, the implication was clear; NeXT became solely a software company, there was no involvement as a hardware company at all. NeXT had become what Microsoft was, and Apple had never been. And the NeXTSTEP operating system would make a lasting impact on computing. The first graphical Internet browser was written for NeXTSTEP, and Apple's future operating systems (including iOS) would be a descendant of NeXTSTEP.

Within his role at NeXT, Jobs showed more precise software and development expertise than he had in the early days at Apple. Deatherage provided an example of one instance where Jobs "followed a visitor to Next

[sic] into the parking lot to argue over an obscure point in Objective-C development—not something many CEOs could have done. But he was not an engineer or programmer at Apple in any significant way."[20]

THE TOOLS TO CREATE SOFTWARE IN THE INTERNET AGE

Bill Gates didn't quite understand what would happen with the Internet in 1995. Steve Jobs almost precisely understood what would happen with the Internet in 1995, and that understanding was one of the reasons his company's work at NeXT was so valuable. Jobs described four major processes that would happen on the Internet in the near future, which would help him in a future initiative.

> People will eventually do four different things on the Web. The first is static publishing, where someone creates a Web page that doesn't change unless they themselves change it. Second, dynamic publishing, where the computer constructs the Web page on the fly based on input from the user and information from a database. A perfect example is the Federal Express Web site, where you type in your package number and it tells you its delivery status with no human intervention. Third, commerce. To do it, you have to hook the Web up to your internal computer systems so you can take orders from the Web. And fourth is internal custom applications, in which a corporation puts its own applications—like a brokerage's program to buy and sell stock—on the Web so that any department, whether it's running Macs or PCs with Windows, can access it. The latter three of those four things are all custom software. We said, that's what we've perfected over the last 10 years—the ability to write custom software 10 times faster.
>
> —Steve Jobs[21]

The idea of routine commerce on the Internet was very innovative. In fact, businesses could not even have a commercial website until the mid-1990s, and few individuals around the world even had access to the Internet.

In an interview with *Wired* magazine in early 1996, Jobs continued to be right about the future of technology and the internet. He foretold that web browsers would become free—in the mid-1990s, some users had to pay for web browser software just to surf the Internet. He accurately spoke about the rise of e-commerce. And he spoke of some of his worries about society and the potential for making the country better as a group (although he had respect for individuals).

> We live in an information economy, but I don't believe we live in an information society. People are thinking less than they used to. It's primarily because of television. People are reading less and they're certainly thinking less. So, I don't see most people using the Web to get more information. We're already in information overload. No matter how much information the Web can dish out, most people get far more information than they can assimilate anyway.
>
> I'm an optimist in the sense that I believe humans are noble and honorable, and some of them are really smart. I have a very optimistic view of individuals. As individuals, people are inherently good. I have a somewhat more pessimistic view of people in groups. And I remain extremely concerned when I see what's happening in our country, which is in many ways the luckiest place in the world. We don't seem to be excited about making our country a better place for our kids.
>
> —Steve Jobs[22]

NEXT IS PURCHASED. BY APPLE.

At the end of 1996, NeXT (a company controlled by Steve Jobs), agreed to be purchased by Apple, for the software the company had developed while attempting to become a hardware producer. That was the end of the first path taken; the second path taken would become Pixar Animation Studios.

Chapter 8

PIXAR AND DISNEY

Pixar is commonly known as the second company that Jobs founded post-Apple. But the company itself wasn't really *founded* by Jobs. The company was an acquisition, with Jobs paying $5 million to acquire the existing workgroup from Lucasfilm. Jobs believed that he could run the struggling division more effectively than it was run within Lucasfilm, and the leadership of the team agreed. The existing leaders of the Lucasfilm division were beneficial to Jobs on a personal level, although many of his later employees may still have disagreed with the assertion:

> From John Lasseter and Ed Catmull at Pixar, Jobs learned to listen to others, temper his abrasiveness and tantrums, and better manage people.[1]

Tempering his abrasiveness and tantrums might be disputed, but he very clearly learned to manage people more effectively. In the 1988 Pixar Business Plan, Jobs was listed as Chairman of the Board with an already impressive pedigree.

> Mr. Steve Jobs co-founded Apple Computer in 1976. He served as Chairman of the Board and Executive Vice President and

> General Manager of the Macintosh Division. Mr. Jobs co-founded NeXT, Inc. in September 1985, where he serves as President and Chairman of the Board. In February of 1986, Mr. Jobs and the employees of Pixar purchased Pixar from Lucasfilm, Ltd.[2]

Dr. Edwin E. Catmull was President, Chief Executive Officer, and Director of Pixar. His appointment as President showed a lot of trust from Jobs; Catmull wrote about Jobs as part of his own memoirs. One of Catmull's early realizations was that Jobs never believed he was wrong when people disagreed with him; Jobs felt his beliefs would always win out if he could "just explain it to them until they understand."[3]

Catmull discussed the initial negotiations for Jobs to purchase Pixar from Lucasfilm in early 1986. The Chief Financial Officer for Lucasfilm decided to intentionally show up late for the negotiations to show he (and Lucasfilm) had more power than Steve Jobs in the negotiations. The CFO didn't get what he expected.

> The morning of the big negotiating session, all of us but the CFO were on time—Steve and his attorney; me, Alvy, and our attorney; Lucasfilm's attorneys; and an investment banker. At precisely 10 A.M., Steve looked around and, finding the CFO missing, started the meeting without him! In one swift move, Steve had not only foiled the CFO's attempt to place himself atop the pecking order, but he had grabbed control of the meeting. This would be the kind of strategic, aggressive play that would define Steve's stewardship of Pixar for years to come—once we joined forces, he became our protector, as fierce on our behalf as he was on his own. In the end, Steve paid $5 million to spin Pixar off of Lucasfilm—and then, after the sale, he agreed to pay another $5 million to fund the company, with 70 percent of the stock going to Steve and 30 percent to the employees.
>
> —Edwin Catmull[4]

Jobs had demonstrated that he was going to take control of the situation with Pixar, and he even designed the entire Pixar facility to meet the mission of the organization:

> Built on the site of a former cannery, Pixar's fifteen-acre campus, just over the Bay Bridge from San Francisco, was designed,

> inside and out, by Steve Jobs. (Its name, in fact, is The Steve Jobs Building.)
>
> —Edwin Catmull[5]

Jobs was a serial entrepreneur—he had invested at least $10 million into Pixar and his firm was working toward releases that would finally give him a positive return on his Pixar investment. Like with the original Apple, Jobs attempted to sell the entire firm three times before finding the largest success.

> Three times between 1987 and 1991, a fed-up Steve Jobs tried to sell Pixar. And yet, despite his frustrations, he could never quite bring himself to part with us. When Microsoft offered $90 million for us, he walked away. Steve wanted $120 million, and felt their offer was not just insulting but proof that they weren't worthy of us. The same thing happened with Alias, the industrial and automotive design software company, and Silicon Graphics. With each suitor, Steve started with a high price and was unwilling to budge. I came to believe that what he was really looking for was not an exit strategy as much as external validation. His reasoning went like this: If Microsoft was willing to go to $90 million, then we must be worth hanging on to.
>
> —Edwin Catmull[6]

One of the Pixar vice presidents commented on changes in Jobs's behavior between the late 1980s and the mid-1990s.

> After the first three words out of your mouth, he'd interrupt you and say, "O.K., here's how I see things." It isn't like that anymore. He listens a lot more, and he's more relaxed, more mature.
>
> —Pamela Kerwin[7]

TOY STORY AND AN INSTANT BILLIONAIRE

After the release of *Toy Story* in 1995 (almost 10 years after Jobs's acquisition), Pixar issued a stock market IPO like Apple did in 1980. And "the stock flew, making Jobs an instant billionaire."[8] The instant billionaire had been twenty years in the making (10 at Apple, almost 10 at Pixar); Jobs's net worth had been estimated at a little over

$100 million when he left Apple in 1985. Ten years later in 1995, he was now a billionaire, theoretically without a need to innovate, work, or save any other division or corporation.

In 1996, Jobs articulated that his team was trying to do something that had only been done previously by Disney. Building a third profitable company was hard enough. Animation and the movie industry was an additional challenge; in some years, animation companies might not have any new releases to drive revenue at all. The first movie release *Toy Story* was nine years and nine months after Jobs's acquisition of the company.

> What we are trying to do ... is build the second great animation studio. And it turns out that's not really an easy thing to do.
>
> —Steve Jobs[9]

That same year, Steve Jobs and Pixar's John Lasseter were invited to an interview on *The Charlie Rose Show* together. Rose asked Jobs a question about Apple during the broadcast, but Jobs delayed responding to his question, until the show was no longer being recorded. Then he made a shocking pronouncement about Apple.

> Before he went back to Apple, we were on *Charlie Rose* together. Charlie asked him about Apple on the air, and Steve didn't really answer. But after the cameras were off, he turned to Charlie and said, "I know how to save Apple. But they're just not listening to me yet."
>
> —John Lasseter[10]

SELLING PIXAR

Pixar had a partnership with Disney for the distribution of films but the partnership was failing in the early 2000s. There were two major parts of the dispute:

- a financial component
- a personality conflict between Steve Jobs and Disney CEO Michael Eisner

Given the personality dispute, there was no quick resolution forthcoming but Jobs was not going to lose.

> The relationship went sour when Michael didn't treat Jobs and the Pixar machine as a giant creative engine, he treated them as second-class citizens.
>
> —former Disney board member Stanley Gold[11]

> Pixar put a deal on the table that was almost insulting to Disney. … It seemed like Steve Jobs wanted to part ways with Disney.
>
> —Jordan Rohan[12]

Disney had a problem, though. Starting with the release of *Toy Story* in 1995, the films with new characters that were popular with children and parents—and the films that were creating sequels—were not made by Walt Disney Feature Animation but by Pixar Animation Studios. After Eisner's departure from Disney, Jobs allowed Pixar to negotiate once again with Disney's new CEO Robert Iger (who had noticed that the most beloved new "Disney" characters were actually Pixar characters).

By early 2006, the company that had created *Toy Story* and sequels, *A Bug's Life*, *Monsters, Inc.*, *Finding Nemo*, and *The Incredibles* had found an organization to purchase Pixar but allow the animation division to continue work on movies. As Jobs had done less than a decade previously with NeXT, Pixar was also sold to a firm that was facing challenges. The year 2006 saw a confluence of factors where Pixar and Disney's existing contract was ending, Disney was the only other firm consistently known for animation, and Disney realized that the characters created in the most recent decade were Pixar's.

> As we approached the end of our relationship with Disney and we looked at our future, we were at a fork in the road … Disney is the only company with animation in their DNA.
>
> —Steve Jobs[13]

This was the end of the second road taken; after coming to another fork, Jobs could have elected to take yet another road—perhaps an independent one with Pixar—but instead sold the company to Disney. For Jobs, the sale of Pixar made him the single largest shareholder of Disney, as he received $3.7 billion worth of stock and a seat on the Board of Directors. And the sales of NeXT and Pixar left only one corporate focus in Jobs's mind: Apple.

PART 4

SECOND RUN AT APPLE

Apple doesn't need a CEO, they need a messiah.

—Jean-Louis Gassee[1]

Apple had bought NeXT for access to the NeXTSTEP operating system; there were effectively no other alternatives.

MacWorld once claimed, "if Steve Jobs had never returned to Apple after 1985, he'd still be remembered for the Macintosh."[2] In fact, Steve Jobs might not be remembered for any Apple products at all if he hadn't come back to Apple; he was personally convinced the firm was on a direct path to failure. So much so that before Apple decided to negotiate with NeXT, Jobs had sold all of his Apple shares from the co-founding.

> Gil Amelio was running the place. So I was also thinking, "Do I really want this $20 million worth of [Apple] stock when I think the company is going to be worthless in a year?" So I sold it.
>
> —Steve Jobs[3]

Chapter 9

SAVING APPLE FROM BANKRUPTCY

APPLE ACQUIRES NEXT

When Jobs first met with Apple about a partnership between NeXT and Apple, he had not been to the Apple campus in more than 11 years; he had never even visited after he left in 1985 following his conflict with Sculley. As part of his meeting and discussion with the newest Apple CEO, Jobs specified that licensing the operating system from NeXT was not the wisest path—if Apple were truly interested in the operating system, they should instead purchase the entire company.

> Jobs arranged to meet with Amelio, Hancock and Doug Solomon, an Apple strategy executive, at Apple's offices in Cupertino on Dec. 2. "It was the first time I had set foot on the Apple campus since I left in 1985," Jobs says.[1]
>
> After the demonstration, as a good faith gesture, Jobs handed over to Apple the financial disclosure documents NeXT had been preparing for its stock sale. And Jobs invited Amelio to his home to get acquainted and discuss strategy. "My advice," Jobs recalls, "was that if Apple was going to go with our technology, they

> should buy the company instead of licensing the software. You need the people for something as vital as an operating system."[2]

Apple took the bait, paying a little more than $427 million for NeXT, a company in which Jobs owned more than half the shares from his initial investment of $10 million. With the purchase, Apple picked up the visionary who had left 11 years ago but had originally been involved in Apple's growth from zero sales to more than a billion dollars annually. Apple got a new operating system and framework for future Apple products that were not yet even on the drawing board.

> For about $400 million, Apple gets a new operating system, fresh engineering blood and a part-time 41-year-old adviser to the chairman who was once the chairman himself. Thus ends Apple's embarrassing public search for its next-generation operating system.[3]

APPLE UNDER GIL AMELIO

Apple was failing under Sculley's successor, Michael Spindler. Spindler offered to sell Apple to a number of potential suitors, including IBM, Sun Microsystems, and Philips. The asking price was not high; the company was almost sold to Sun Microsystems for $6 per share, well less than 1 percent of Apple's adjusted market value today. Years later, the co-founder and long-term CEO of Sun Microsystems made the following candid declaration about the progression of Apple products if Sun had been able to acquire the company, as he was not the visionary Steve Jobs would prove to be:

> If we had bought Apple, there wouldn't have been iPods or iPads. I'd have screwed that up.
>
> —Sun CEO Scott McNealy[4]

In early 1996, Spindler was out and Gil Amelio was in as CEO, but innovation stagnation and continued financial losses had already set the stage for Jobs's full-time return to Apple. Amelio was quickly describing the company's marquee product not in the design terms of

Jobs but in terms of a sturdy product that cost a little more, not concerned at all about rapidly declining market share.

> Apple insists that gobbling up market share is no longer part of its strategy. Instead, the Macintosh will become "the MagLite of computers," Amelio promises, comparing future Apple products to a slightly more expensive, yet durable brand of flashlight.[5]

Tellingly, analysts were not convinced that the approach of Amelio was working. Even worse, Apple had not discussed any new products in the half year since announcing the turnaround.

> I haven't seen the company saver yet ... We need to see the new stuff now.
>
> —John Rossi of Robertson Stephens.[6]

Amelio attempted to improve the company's financial position but had his new advisor by early 1997—Steve Jobs was given a part-time role at Apple as part of the acquisition of NeXT. There was a reason he had already sold his Apple shares; Jobs was convinced Apple would fail under Amelio (conveniently, the sale of NeXT involved cash as well as Apple stock). The restructuring started under Amelio was to include employees and products.

> Amelio promised shareholders that there would, indeed, be cuts, as he and other Apple execs prepare a "hit list" of products (and, presumably, employees) that will have to go ... Our take. Wall Street investors will be eager for a draconian reorganization plan after Apple's $400 million acquisition of NeXT, and layoffs surely must be part of that plan. But across-the-board cuts would be unwise, as Apple has to contend with some other disturbing news: Apple is losing favor with one of its stronghold markets—education. International Data Corporation reports that for the first time ever, elementary schools plan to buy more PCs than Macs.[7]

In a fateful development for Amelio's time as CEO, spending cash on NeXT for Apple's new operating system exacerbated the problems with cash flow. Customers were not buying Apple's products; in the previous year alone, Apple had fallen from 8 percent of the global market share to 5 percent. That was a precipitous drop in market share in a year the use of computers was rapidly increasing. Apple was simultaneously facing problems in "cash shortages, faulty products, delays in upgrading its Macintosh operating system, and an overloaded product line."[8]

The situation at Apple appeared to deteriorate further over the year, with analysts believing Apple had actually regressed between 1996 and 1997. The market share for Apple products continued to decline.

> Amelio also discussed Oracle Chairman Larry Ellison's interest in possibly buying the company to make Apple-branded network computers ... Analysts were not buying what Amelio was stating about Apple ... "The rest of the industry is dramatically lowering the cost of managing PCs, working on appliance like PCs and NetPCs," said Rob Enderle, a senior analyst at Giga Information Group, in Santa Clara, Calif. "I don't see Apple doing any of that. It looks like things are going to continue to degrade."[9]

Independent observers didn't believe what Amelio was saying, experts didn't believe Apple was following market trends, and one of Steve Jobs's best friends was potentially speaking about purchasing the firm. And in speaking to that best friend, Amelio ostensibly made a comment resembling the following to Larry Ellison at a cocktail party:

> Apple is a boat. There's a hole in the boat, and it's taking on water. But there's also a treasure on board. And the problem is, everyone on board is rowing in different directions, so the boat is just standing still. My job is to get everyone rowing in the same direction so we can save the treasure.
>
> —Gil Amelio

> But what about the hole?
>
> —Larry Ellison, 1997[10]

By this point, Amelio likely knew (or should have known) his tenure at Apple was coming to an end. In July 1997, Amelio was terminated from Apple. The next year, he released his book *On the Firing Line*. In the book:

> Larry Ellison, the billionaire founder of Oracle, comes off as "conniving" and "manipulative." But Amelio reserves his harshest judgment for Steve Jobs. Amelio says Jobs's "technical understanding only goes a micron deep." Most of all, though, Amelio feels that Jobs sold him down the river, sucking up to him in person while manipulating the press and maneuvering the board into firing him.[11]

Steve Jobs—while serving in an advisory capacity back at Apple—had already made a mark in terms of entitled behavior. He was using the disabled parking spaces (and receiving citations) so frequently that Apple removed the disabled parking spaces in order to placate Jobs.

> Jobs insists on parking in the most convenient spots on the Apple campus: those by the door reserved for the disabled. He's run up so many parking fines, which Apple pays, that the company has painted over the universal blue sign so Jobs can park guilt- and ticket-free.[12]

But he was back at Apple, and he had no CEO in the way because there was no CEO (at all) originally. Jobs then became the interim CEO, crediting his wife for helping him stay at Apple while describing the situation as:

> Much worse than I could imagine. The people had been told they were losers for so long they were on the verge of giving up. The first six months were very bleak, and at times I got close to throwing in the towel too.
>
> —Steve Jobs[13]

MOST ADMIRED ENTREPRENEURS

At the 1997 *Inc.* 500 Conference, participants were asked to identify their most-admired entrepreneurs. Steve Jobs ranked third, behind

Bill Gates (who received the majority of votes as a co-founder of Microsoft and Corbis) and Ted Turner (founder of television stations including CNN and TBS and former owner of the Atlanta Braves).[14] The very interesting part of these results was that Jobs's true resurgence at Apple had yet to begin; he was just the interim CEO, and few new the full power of the software he had brought over from NeXT.

DOESN'T WANT TO BE CEO

Upon Amelio's ouster, Jobs repeatedly stressed that he did not want to be CEO of Apple and was searching for Apple's next CEO. The media was similarly interested in whether Jobs would take the post, while Jobs was preoccupied stacking the Board of Directors with industry titans (who also happened to be personal friends).

JULY 1997

> *[Jobs] has indicated he does not want to be considered for the CEO job.*
>
> —Edgar Woolard[15]

Observers from publications such as *Newsweek* claimed Jobs would be the best possible CEO for Apple at that point in time.

> And Jobs has already set the stage. His rousing speeches to developers at recent conferences, his forceful articulation of Apple's problems and his proposed solutions have already impressed the board and infused the company's battered troops with hope. Everyone agrees that Apple needs a charismatic, visionary leader, supremely confident in his ability to convince the computing world that Apple has a future. At that, no one would be better than Jobs.[16]

AUGUST 1997

A newspaper reported that Jobs would take the CEO position, "according to a published interview with [Larry] Ellison in the French financial newspaper *La Tribune* last week."[17] *Jobs* had to deny again—to the staff at Pixar and to the media—that he was a candidate to be the CEO of Apple.

After reports last week that he'd take the post, Jobs reassured employees at Pixar, his film-animation firm, that he had declined the offer.[18]

My name is not in the hat to be CEO.—Steve Jobs[19]

Neither did Apple outline a coherent plan for its future. That task may be up to a new CEO. Jobs, who doesn't want the job... [20]

NOVEMBER 1997

Jobs was interviewed by the *San Jose Mercury News*; the interviewer pressed Jobs on the CEO search. Jobs suggested a CEO would be placed by early 1998.

Reporter: What about the CEO search?
Steve Jobs: It's happening. The same story. We're looking, we've got some pretty healthy candidates. We hope to have somebody in the saddle by very early next year.
Reporter: Could it possibly be you?
Steve Jobs: I've been very consistent on this.
Reporter: Could you state the answer for the record?
Steve Jobs: For the record, we're looking for a CEO and expect to have someone in the saddle early next year. I'm out interviewing people now.[21]

While Apple didn't have a permanent CEO, Jobs was busy stacking the Board of Directors with his friends. The billionaires Steve Jobs added to the Board of Directors would make the life of any ordinary CEO reporting to that Board of Directors a challenge.

Larry Ellison, software's second-leading billionaire. Steve Jobs. Jerry York, the turnaround king from Chrysler and IBM. And on the bench, shepherding his new $150 million investment in Apple: William H. Gates, probably the smartest and hardest techno-player of them all. There's just one problem: these

> strong-willed, charismatic tycoons rarely agree on anything. They've devoted their lives to beating each other up. So now they're supposed to make happy-happy at Apple?
>
> My prediction is that Jobs will ultimately be named chairman, and Apple's board will be running the business.
>
> —Geoffrey Champion, head of a search firm[22]

The willingness of these individuals to serve on the Board of Directors showed a commitment to Apple, a firm that would have to make many tough decisions to put itself back on the path to recovery. The exceptionally strong Board of Directors could be a blessing or a curse for Apple. Clearly, the Board of Directors was one of the strongest technology boards ever created.

> "Apple needed some heavy hitters on their board," explained the spokesperson from CalPERS. "It was an improvement to their board structure." Kenny feels that Ellison's election also makes it easier for Apple to attract a new CEO. "It gives Apple enormous credibility."[23]

The search for a permanent CEO did not progress rapidly; perhaps it was hard for Apple to attract a new CEO after all. It took two-and-a-half years before Steve Jobs finally selected and announced Apple's new permanent CEO.[24] There was no change of leadership when the announcement finally came in January 2000, as the permanent CEO was Steve Jobs.

MAKING PEACE WITH MICROSOFT

> *Apple didn't have to beat Microsoft. It had to remember what Apple was. Microsoft was the biggest software developer around, and Apple was weak. So I called Bill up.*
>
> —Steve Jobs[25]

So what was this money from Microsoft, and why had Jobs called his friend Bill Gates? Given the consistent declines at Apple over the past few years and financial losses (including $1.5 billion over the previous year-and-a -half), the future of Apple was dire:

> Apple has some tremendous assets, but I believe without some attention, the company could, could, could—I'm searching for the right word—could, could … die.
>
> —Steve Jobs[26]

Jobs made a presentation at a major Apple event. He was joined on-stage—via video—by at least one individual who was viewed unfavorably by many Apple users.

> This era of competition between Apple and Microsoft is over, as far as I'm concerned … This is about getting Apple healthy.
>
> —Steve Jobs[27]

After extensive preparation, including getting a video from friend Larry Ellison and coordinating with Bill Gates as to the clothing that would be worn, he thanked Bill Gates for the support, which would be Apple's best, last chance of success.[28] Without raising external funding, there was a high probability Apple would be seeking bankruptcy in the very near future. But a partnership between Steve Jobs and Bill Gates was still more than many Apple customers could accept.

Reporter Rick Webb would later write of the event that there were now tones of George Orwell's *1984* between Microsoft (Big Brother) and Apple, just as Apple had claimed between IBM (Big Brother) and Apple when the Macintosh was released. He commented: "In 1997 in Boston I had the pleasure of witnessing in person what Steve Jobs called 'my worst and stupidest staging event ever.' Onstage at Macworld Boston, Jobs announced his settlement of legal disputes and a partnership with Microsoft. And in a move eerily reminiscent of his landmark 1984 advertisement, Bill Gates' satellite-broadcast image filled the hall, a looming face looking every bit the overlord out of place. He was the Orwellian big brother we had come to despise. People booed as he spoke. In the end, the deal was probably a good thing, but the symbolism was catastrophic."[29]

> I'm sure some people want to cling to old identities. I was a little disappointed at the unprofessional reaction. On the one hand,

> people are dying to get the latest release of Microsoft Office on their Macs, and on the other hand, they're booing the CEO of the company that puts it out. It seems really stupid to me ... Apple has to move beyond the point of view that for Apple to win, Microsoft has to lose.
>
> —Steve Jobs[30]

Microsoft would soon be fighting the U.S. Department of Justice on alleged violations of the Sherman Antitrust Act, so helping a competitor would put the firm in a good light. According to the corporate release from Microsoft at the time:

- Microsoft will develop and ship future versions of its popular Microsoft Office productivity suite, Internet Explorer and other Microsoft tools for the Mac platform.
- Apple will bundle the Microsoft Internet Explorer browser with the Mac OS, making it the default browser in future operating system software releases.
- The companies agreed to a broad patent cross-licensing agreement. It paves the way for the two companies to work more closely on leading-edge technologies for the Mac platform.
- Apple and Microsoft plan to collaborate on technology to ensure compatibility between their respective Virtual Machines for Java and other programming languages.
- To further support its relationship with Apple, Microsoft will invest $150 million in non-voting Apple stock.[31]

The settlement of legal disputes mentioned by Webb was missed by many in the audience who were still stunned by the Jobs–Gates friendship and joint appearance. A purchase of $150 million in non-voting stock only does so much for a firm that has been losing over $80 million a month for the last year-and-a-half. In addition to the $150 million and technical collaboration, "Jobs pointed out that people had been so shocked they missed the big news: Microsoft would be paying an undisclosed amount to settle claims that it had used seminal Apple computer patents."[32] The arrangement was not only a good thing for Apple, investors knew it was a very good thing. The stock price rose by more than one-third that day alone.

WHY DID JOBS CALL MICROSOFT?

In 1983, Apple sold 11.2 percent of all computers sold in the world. By 1997, that figure had fallen to 3.3 percent. A firm that had once sold one in nine computers was now selling one of every 30.[33]

> Apple will use Microsoft's money to continue developing new products for graphics, publishing, and education. Apple still dominates in these areas; by Jobs's reckoning, Macs account for more than 80 percent of computers used in graphic design and about 60 percent of computers in schools. And some predict that Microsoft's software commitment may lead to even better news for Mac users in a wider range of software choices.[34]

Jobs's assessment there was partially correct while tellingly incorrect. In terms of software sales at the time, Macintosh had a substantial share of the market in graphics:

Desktop publishing software: 62.4 percent
Presentation software: 51.6 percent
Drawing and painting software: 43.9 percent

Apple was still the largest vendor when it came to computers sold to educational institutions in the United States. These purchasing characteristics tell everything that's necessary to understand the failure of Apple.

Apple	29.6 percent
Dell	9.6 percent
Compaq	9.1 percent
IBM	7.3 percent
Gateway 2000	6.6 percent[35]

The reason for the high use of Macintosh in graphics was due to large purchases of Apple products in the past; their market share in graphics was due to users on older computers, not new computer sales. Aside from the 30 percent market share of products to schools, Apple was effectively selling no computers to anyone else. That's how Apple had fallen to selling just one of every 30 computers worldwide. Take away educational sales and Apple would have sold far fewer than one of every 30 (and already have failed).

In an interview, Jobs recognized that Apple had a number of strategic mistakes already in progress before Gil Amelio was removed. When questioned later that year why Apple was announcing a series of new computers where the cheapest option was $2,000:

> You're exactly right. The interesting thing is that Apple has lost the focus on products at the low end of the consumer price points and I think that's a mistake.
>
> —Steve Jobs[36]

In a tactful way, Jobs described the process by which he had to remove individuals from Apple. Those who were removed were the ones who were "not so outstanding" and would be the most likely to speak poorly of Jobs in the future.

> I've discovered there are really some outstanding people at Apple. I've also discovered there are some who have not been outstanding. And I've been trying to move the ones that are not so outstanding out of the way and move the ones that are outstanding into key jobs. So some of the people who have been moved out of the way are going to say terrible things about me probably, but that goes with the territory. There's clearly needed to be some cleaning house at Apple. It's mostly behind us.
>
> —Steve Jobs[37]

RETURN OF CHIAT\DAY

Remember when Steve Jobs—in 1985—purchased an advertisement telling the fired advertising agency CHIAT\DAY there was life after Apple? CHIAT\DAY (by then TBWA\CHIAT\DAY) still had another life with Apple remaining. In fact, Apple's existing advertising company resigned as soon as they became aware Jobs was leading a review of the advertising program.

> Mr. Jobs's decision to become, by default, the most influential member of the review committee spins the search in a new

> direction. Apple's decision last month to review prompted BBDO West, Los Angeles, to resign.[38]

A few months later, Jobs removed his internal team leading advertising, brought in a hand-selected person to run the division (Allen Olivo), and spoke not only of maintaining advertising spending at $100 million a year, but spending that money more effectively. And one of those ways of spending funds more effectively was in the "Think Different" advertising campaign.[39]

THINK DIFFERENT

CHIAT\DAY's "1984" ad for Apple was targeted at the established IBM. Their next advertising campaign for Apple was similarly targeted at IBM. IBM was using the line "I think therefore IBM." Apple's new (old) advertising agency established a campaign to instead "Think Different," which was intentionally made non-grammatical ("different" rather than "differently," reinforcing the concept of "Think Different").

> IBM's ... verbal identity has changed subtly too, perhaps exemplified by the advertising line "I think therefore IBM." IBM's language has become less technologically obsessed, less to do with bits and bytes and more to do with having an interesting way of thinking. Indeed, in terms of verbal identity, it drew the response from Apple "Think different," which was a blow aimed at IBM's supposed weak spot: its association with "blue suit" conformity.[40]
>
> Apple's "think different" (the advertising articulation of its brand idea) demonstrates an insight that its target customers believe in and desire a better way for humans to interact with technology.[41]

WHAT JOBS DID RIGHT THE SECOND TIME

Benj Edwards claimed Steve Jobs made seven key decisions upon returning to Apple. With more than a decade away from the firm, he was clearly a different executive—and individual—than the one who had been pushed out of the company in 1985. He also exhibited a far better idea of business concepts and saw failures that had continued from Sculley's turn as CEO to Amelio's turn as CEO.

The seven concepts Edwards articulated were: Taking the Reins; Trimming the Fat; Cleaning House; Plugging Leaks; Burying the Hatchet; Killing the Clones; and Trusting Jonathan Ive. These can all readily be seen as positive strategic steps Jobs took upon returning, although Macintosh clones and Jonathan Ive were not part of Apple when Jobs departed in 1985.[42]

Taking the Reins—Under Sculley's leadership, Jobs was marginalized, and then displaced, from the firm. Under Amelio's leadership, Jobs similarly turned the Board of Directors against Amelio, which allowed him to return to the CEO position.

Trimming the Fat—Amelio had frequently spoken about a need to shut down some of the Apple product lines and had indeed begun to do so. After Jobs's return, the firm focused on doing a few things well rather than be involved in a broad array of ancillary products and services.

Cleaning House—Jobs used the Board of Directors to remove Amelio; he then led a period of Board renewal, where the Directors (who are tasked with overseeing the executives including Jobs) were themselves removed and replaced by individuals friendly to Jobs. This was one way he ensured he had allies helping to control the firm's future direction.

Plugging the Leaks—Over the next 14 years, Jobs became well known for the press conferences where new, often revolutionary products were released. In order to make these large pronouncements, employees were no longer allowed to speak to the media in advance of major new products. While new products were often rumored, everyone heard about the newest products (and features) at the same time.

Burying the Hatchet—Jobs realized that Apple, which had shrunk to approximately 3 percent of the global computing market, was not going to compete for the majority of the market with PCs running Microsoft Windows. The alternative was to instead focus on Apple's core competence and selectively partner with Microsoft, as the firm had done in Jobs's initial period with Apple (the period with runaway growth).

Killing the Clones—Apple had been allowing other computer manufacturers to license the Apple Operating System (OS7). However, this had a detrimental effect. While Microsoft was selling their Windows operating system to almost every manufacturer throughout the world, Apple's licensing of their operating system took away from sales of their most profitable product; the hardware. Microsoft did not make computers, so their financial interest was in creating an environment where computer manufacturers licensed their software. Apple licensing software inadvertently gave customers the ability to buy a less expensive computer, harming Apple's own computer sales.

Trusting Jonathan Ive—Jonathan (Jony) Ive was a design expert who—like Jobs—may have had ideas the previously leadership did not—or could not—appreciate.

END OF THE CLONES

Clones for the Apple OS were initially proposed by one of Steve Jobs's friends. Bill Gates proposed the idea to John Sculley in the year 1985, the same year of Jobs's initial departure.

> Apple should license Macintosh technology to three-five significant manufacturers for the development of Mac Compatibles ... Microsoft is very willing to help Apple implement this strategy.
>
> —Bill Gates[43]

However, Apple still didn't realize the scope that IBM compatibles (first using DOS, then using Windows) would quickly take. If Apple had moved to compatibles early, Windows might be the secondary operating system in the world (or perhaps may not even exist at all). Apple didn't move quickly—it took nine years of failed innovation without Jobs before Sculley's successor Spindler allowed clones. Jobs quickly put an end to the clones. Given Spindler and Amelio both had short tenures at Apple, some manufacturers were exceptionally upset at Jobs for eliminating the clones so early after Apple started the program.

JONATHAN (JONY) IVE

"(Expletive), you've not been very effective, have you?"

—Steve Jobs to Jony Ive[44]

Jonathan (Jony) Ive was instrumental in designing many of the Apple products used today; in fact, he was knighted by Queen Elizabeth II of England as a Knight Commander of the British Empire for his design work.[45] Jobs didn't require Ive to conduct any public speaking, because Ive didn't wish to be involved in public speaking.

> I'm shy ... I'm always focused on the actual work, and I think that's a much more succinct way to describe what you care about than any speech I could ever make.
>
> —Jony Ive[46]

Ive was almost fired when Jobs returned to Apple. Ive had a resignation letter written, and Jobs was considering alternatives. Then Jobs visited the design studio and exclaimed something to the effect of "(Expletive), you've not been very effective, have you?" The work in the design studio was worthwhile but no one had the voice to declare the studio's value. Jobs was energized:

> That day, according to Ive, they started collaborating on what became the iMac. Soon afterward, Apple launched its "Think Different" campaign, and Ive took it as a reminder of the importance of "not being apologetic, not defining a way of being in response to what Dell just did." He went on, "My intuition's good, but my ability to articulate what I feel was not very good—and remains not very good, frustratingly. And that's what's hard, with Steve not being here now."[47]

CONTROLLING A LEAKY SHIP

Upon his return to Apple, Jobs commented in 1997 "Wouldn't it be funny, a ship that leaks from the top?" when talking about a lack of product announcements that day. While not picked up at the time, the single line had two profound meanings.

- The first meaning was a not-so-veiled insult of Gil Amelio and the "leaking ship" metaphor, four months after his termination.
- The second meaning was a philosophical framework where Apple's major product announcements are tightly guarded, released only by the CEO.

Shortly thereafter, Jobs would begin making the pronouncements of all new products on behalf of Apple, very tightly restricting the ability of staff to talk about products in advance to the media. There would be no leaks, under threat of employee termination or lawsuits. As Seybold noted immediately after the conference:

> Without any big announcements of products or policy ("Wouldn't it be funny, a ship that leaks from the top?" he asked), Jobs used his opportunity to convince the audience that much was being done and it would be sufficient to bring Apple and the Macintosh through its current crisis. The thrust of his message was the 11 things he and the company already have done, including appoint a new board of directors, beef up Apple's distribution, recommit to its developers and improve its communication with users.[48]

Leander Kahney later declared the combination of secrecy and high profile announcements to be one of the 10 Commandments of Steve, noting "Nobody at Apple talks. Everything is on a need-to-know basis, with the company divided into discrete cells. The secrecy allows Jobs to generate frenzied interest for his surprise product demonstrations, and the resulting headlines ensure lines around the block."[49]

1998—FIRST FULL YEAR BACK AT THE HELM

Jobs had made it through a partial year heading Apple but there were still a lot of items to fix. Multiple executives said it was easier to meet with Jobs—in terms of personality—on Fridays:

> We were always happy when we had a Friday meeting with Steve . . . because Friday was the day he was at Pixar, and he was always in a good mood there.
>
> —Andy Dreyfus[50]

Seybold described Jobs's presentation at their publishing conference as an hour-long commercial. Despite many years of failures at Apple, "the fact that the Seybold audience loved every minute of it proves that Apple still holds a privileged position in the publishing industry."[51] Jobs took charge, following up by noting the lack of software programs for the Macintosh was actually Apple's fault. The software developers would have handled the sales declines but were burned in their interactions with the company itself.

> Let's focus first on the real problem. Forget the perception. The real problem is that a lot of developers have had a really tough time dealing with Apple over the last few years. I talked to these folks. It wasn't even about the volume of Mac sales declining—it was problems in dealing with Apple. We fixed almost all of that. The developers are coming back, and it feels really good. We haven't brought everybody back yet, but a lot of them.
>
> —Steve Jobs[52]

Jobs was also able to recruit—and hire—the individual who would eventually take over as CEO more than a decade later. At the initial meeting between Jobs and Tim Cook, Cook was shocked when Jobs told him that he wants to take a strategy targeting individual consumers, while market competitors feared that individuals were no longer profitable. Unlike his hire of Sculley, Jobs did not regret this choice.

> Well I'm just thinking I'm going to meet him and all of a sudden he's talking about his strategy and his vision, and what he was doing was going 100 percent into consumer. When everybody else in the industry had decided you couldn't make any money on consumers so they were headed to services and storage and enterprise. And I thought, I'd always thought that following the herd was not a good thing, that it was a terrible thing to do right? You're either going to lose big, or lose, but those are the two options. He was doing something totally different.
>
> —Tim Cook[53]

CLAIMING THE HIGH GROUND AND FOCUSING APPLE

He used his sell-off of Apple shares under Amelio's leadership in an attempt to claim moral authority; he could craft a compelling narrative that he had returned to fix Apple, not to gather additional riches for himself.

> You could argue, as you did, that I don't have a stake in Apple. But I was able to walk in with some moral authority and say, "Look, this isn't about me or the money I'm going to make. This is about what's right for Apple." It was purer in some ways.
>
> —Steve Jobs[54]

When asked if his management style could be described as persistence, Jobs disagreed. He provided an example from his childhood of a neighbor who was persistent but unwise, saying that one instead has to be fully dedicated to the most important tasks, a lesson he was reinstilling in Apple.

> I don't think of it as persistence at all. When I was growing up, a guy across the street had a Volkswagen Bug. He really wanted to make it into a Porsche. He spent all his spare money and time accessorizing this VW, making it look and sound loud. By the time he was done, he did not have a Porsche. He had a loud, ugly VW. You've got to be careful choosing what you're going to do. Once you pick something you really care about, and it's a worthwhile thing to do, then you can kind of forget about it and just work at it.
>
> —Steve Jobs[55]

The drastic turnaround process had yielded personal drama for Jobs, who received death threats after eliminating the possibility of cloning Apple computers.

> In hindsight that looks smart, but have you ever gotten death threats? That was scary.
>
> —Steve Jobs[56]

1999—THE TURNAROUND IS APPARENT

Jobs was forcefully articulating his full vision for the turnaround of Apple, a company that was building computers for average people, not unaffordable $2,000 basic models.

> Who is Apple? Why is Apple here? Remember, the roots of Apple were to build computers for people, not for corporations. At the time we started Apple, IBM built computers for corporations. Now it's Microsoft and Intel. But there was nobody building a computer for people. Funny enough, 20 years after we started Apple, there was nobody building computers for people again. You know? They were trying to sell consumers last year's corporate computers. We said, "Well, these are our roots. This is why we're here. The world doesn't need another Dell or Compaq. They need an Apple." In the case of Apple, we're going to make it easy as possible to use this.
>
> —Steve Jobs[57]

Jobs also refers to his structure removing politics from the corporation, trusting three key individuals to run three of the divisions. And he reiterates that believing in the consumer mission was a strong point for Apple.

> Number one, everybody is compensated like a startup. Number two, we have a very simple, clear organization. It's very easy to know who has authority for what, who has responsibility for what. There's no politics about it, they're virtually politics-free organizations. There's no turf wars. Avi runs software. John runs hardware. Mitch runs Sales. It's really simple. Number three, we have a very simple mission. It's very easy to communicate what we're trying to do.
>
> —Steve Jobs[58]

As the 1990s closed, observers began to notice that Apple under Jobs was able to sell items effectively based upon color alone—the obsessive nature he had shown in relation to colors in the Apple II and NeXT was paying off.

> But he can sell a computer like the iMac almost entirely on the basis of color. Which is not to say there isn't some pretty cool stuff going on inside there. But forget content of character; it's

> those fun, fruity colors—strawberry, tangerine, lime, grape, and blueberry—and those TV ads with iMacs spinning to the tune of the Rolling Stones' "She's A Rainbow" (totally trumping Bill Gates and his "Start Me Up" Windows 95 campaign) that gets these machines dancing off the shelves.[59]

Within two years of Jobs returning to Apple, the firm had almost quadrupled market share from approximately 3 percent of retail sales to 12 percent of retail sales.[60] Steven Berglas even suggested that this juncture in time would be the perfect time for Apple to remove Jobs; he would have trouble finding a new position in a traditional firm and have to find a new business that was so beleaguered that a hiring of Jobs to bring credibility would be reasonable.[61]

EARLY PORTRAYAL IN THE MEDIA

The book *Fire in the Valley* was made into the movie *Pirates of Silicon Valley* in 1999, with actor Noah Wyle portraying Jobs. Wyle received an unexpected call on his private phone line the next day from someone with a partial compliment.

> I'm just calling to tell you I thought you did a good job. I hated the movie, I hated the script, I think if you had spent a little more time and a little more money and maybe a little more attention to detail, you could have had something there. But you were good.
>
> —Steve Jobs to Noah Wyle, in a phone call the day after *Pirates of Silicon Valley* aired[62]

SEEING THE FUTURE IN 1999

In 1999, *PC Week* commemorated their 15th anniversary by naming a list of the 15 most influential individuals (all male) in the technology industry in the preceding 15 years (1984–1999), with Linda Bridges asking about their accomplishments and also their perception of where the industry would be 15 years later (in 2014). CEOs and founders of 13 of those companies provided input for the article. The two most influential who declined to participate for the article? Steve Jobs and his friend Larry Ellison did not respond, while other friends (such as richest man in the world Bill Gates) did participate.[63]

While Jobs refused to participate in *PC Week*'s commemorative edition, he did tell about what the future would look like in 2014, although he would not live to that year. He spoke about getting a high-speed line connected to his home that would help him connect with his work—whether at home, Apple, or Pixar—as that would one day become standard.

> I'm actually getting ready to put a 45 mg fiber to my house, because I want to find out what that will be like, because everybody's going to have that someday. But I have a pretty sophisticated setup; whether I'm at Apple or at Pixar or at my home, I log in and my whole world shows up on any of those computers. It's all kept on a server. So I carry none of it with me, but wherever I am, my complete world shows up, all my files. Everything. And I have high-speed access to all of it. So my office is at home too.
>
> —Steve Jobs[64]

Jobs had another revelation about the Internet, as web browsers and standards had advanced to the point that webpages would forevermore always look substantively the same regardless of which computer or browser was used.

> No website knows whether it's a Mac or Windows on the other end of the line.
>
> —Steve Jobs[65]

Although Jobs was still a few months from naming himself as the permanent CEO of Apple, he revealed that his leadership of Apple would not end in the near future. He did reveal that at some point in the next 10–20 years:

> I will not be running Apple.
>
> —Steve Jobs[66]

An observer at that time would likely understand that the interim CEO was indeed permanent, and would lead Apple for approximately another decade.

THE LAST YEAR OF THE MILLENNIUM

As the year 2000 wound down and the twentieth century came to an end, Jobs was ecstatic that the work done at NeXT—for major corporations and large universities—was finally getting used on a daily basis for consumers, the market Jobs wanted to serve all along. He had found a forceful point in interviews to stress that Apple created products for individuals, not businesses.

> The thing about NeXT was that we produced something that was truly brilliant for an audience that our heart really wasn't into selling to—namely, the enterprise. I suppose if you were writing a book, this would be a great plot line, because the whole thing circles back. All of a sudden, it's coming out for the market that we would've liked to create it for in the first place—i.e., consumers. So it's a good ending.
>
> —Steve Jobs[67]

At that point in time, Jobs probably had an idea of products that would be the "next Macintosh." Yet even questioned directly, he would not release that information, saying he didn't care about the next Macintosh. He simply wanted to create a company that could (and would) take advantage of any opportunities that arose, unlike Xerox (which had inspired many Apple products but couldn't execute on their ideas).

> People are always asking, "What will be the next Macintosh?" My answer still is "I don't know and I don't care."
>
> Everybody at Apple has been working really hard the last two-and-a-half years to reinvent this company. We've made tremendous progress. My goal has been to get Apple healthy enough so that if we do figure out the next big thing, we can seize the moment. Getting a company healthy doesn't happen overnight. You have to rebuild some organizations, clean up others that don't make sense, and build up new engineering capabilities.
>
> —Steve Jobs[68]

SALARY WAS $1, BUT THERE WERE STOCK OPTIONS, AND AN AIRPLANE …

While Jobs was well-publicized as having a salary of just $1 per year to keep his family on the health insurance coverage, that does not mean that the firm did not spend extensively to keep Jobs as CEO. In fact, there was a modest financial clamor in 2000 about how the firm reported the "gift" of a $90 million private jet to the CEO—whether the item should be considered to be a "recurring" expense in terms of Apple's accounting or recorded as a one-time expense. As one analyst said of the uproar:

> They won't be doing this again … How would you top it? Give him an island? A small country?
>
> —Lou Mazzucchelli, Gerard Klauer Mattison & Co.[69]

Despite divergent paths after the earliest days of Apple, Wozniak responded to a fan when asked if he would fly with Jobs on the new jet at some point in the future:

> Of course, but it's not a desire of mine to add even a single trip to my life.
>
> —Steve Wozniak[70]

Chapter 10

APPLE'S RESHAPING FOR THE MODERN ERA

2001—THE BEGINNING OF THE IERA

As the new millennium began, Jobs realized that if Apple was dedicated to reaching individual consumers, Apple should focus on reaching individual consumers. This is when Apple became the company we know today. First came the showplace for all current and future products: the Apple Stores opening in 2001 provided a demonstration area for customers.

> People don't just want to buy personal computers anymore ... They want to know what they can do with them. And we're going to show people exactly that.
>
> —Steve Jobs[1]

Jobs's references to "personal computers" when talking about the Apple Store? He was exceptionally careful when choosing words, as the Apple iPod would be released five months later. At the 2001 introduction of the first iPod using iTunes, he was confident the product would be a success and not a speculative product.

> Music's a part of everyone's life ... Music's been around forever. This is not a speculative market. And because it's a part of everyone's life, it's a very large target market all around the world.
>
> —Steve Jobs[2]

The iPod was not universally well received by competitors. Microsoft's Bill Gates was not impressed and thought consumers would want many different types of devices to use just for music. Shortly thereafter, Microsoft would try to issue a product that offered price, performance, and capabilities.

> There's nothing that the iPod does that I say, "Oh, wow, I don't think we can do that." There's often, early in the new market, a few products that help get the category to critical mass. In the long run, people are going to buy what gives them the right price, performance, and capabilities. And does everybody want to have exactly the same thing? Probably not.
>
> —Bill Gates[3]

2002—INVESTING WHEN OTHERS WOULD NOT

Like his hero Dr. Edwin Land, a co-founder of former industrial giant Polaroid, Jobs knew the survival of his firm was based upon innovation. While many companies in financial crises look for cost-savings, "Apple under Jobs upped its research-and-development spending, helping the company produce a strong product lineup that could weather tough times."[4]

> The way we're going to survive is to innovate our way out of this.
>
> —Steve Jobs[5]

2003—SUBSCRIPTIONS FOR MUSIC WILL FAIL

While promoting iTunes (and trying to block companies such as MusicMatch), Jobs declared that there was no future for companies who provided music via subscription to consumers.

> *Interviewer*: What about subscription services?
> *Steve Jobs*: Well, they've failed. They've completely failed. Nobody wants to rent their music. They have hardly any subscribers.[6]

MusicMatch was bought out, failed, and then discontinued. But Jobs was only correct at that single point in time. Apple would later purchase Beats Music—a subscription service co-founded by rapper Dr. Dre—after the death of Jobs. Beats Music was incorporated into the Apple Music service, one of many products or services that Jobs declared as unnecessary that were introduced by Apple soon after his death.

THE *ROLLING STONE* INTERVIEW

In his 2004 *Rolling Stone* interview, Jobs—the spokesperson of the iPod and iTunes—stated that record companies were naïve to believe that digital content could be routinely protected. He also came out as an advocate for intellectual property, saying that it is always corrosive to character to steal, although stealing was fine in service to Krishna in terms of situational ethics.

> People need to have the incentive so that if they invest and succeed, they can make a fair profit. But on another level entirely, it's just wrong to steal. Or let's put it this way: It is corrosive to one's character to steal. We want to provide a legal alternative.
>
> —Steve Jobs[7]

Jobs also claimed that there would not be an iTunes movie store, despite his insistence just a few years earlier that the broadband internet he installed at home would become a standard for other home users.

> We don't think that's what people want. A movie takes forever to download—there's no instant gratification.
>
> —Steve Jobs[8]

With more users possessing high speed internet like Jobs installed in 1999, iTunes did begin offering full-length movies. Apple began offering video less than two years later and customers did want the ability to download a movie. The company sold more than a million videos in the first 20 days of offering the option through iTunes.[9]

FIRST DISCLOSURE OF ILLNESS

At the time Jobs gave the *Rolling Stone* interview, he was already ill but hiding a diagnosis of pancreatic cancer from all but his closest family and friends. Yet for most of a year, he tried alternative medicine instead of surgery or other conventional methods.

> His early decision to put off surgery and rely instead on fruit juices, acupuncture, herbal remedies and other treatments—some of which he found on the Internet—infuriated and distressed his family, friends and physicians.[10]

Eventually, he had surgery to remove the tumor in the summer of 2004, then provided the first public acknowledgement that he had cancer in a communication to employees.

> This weekend I underwent a successful surgery to remove a cancerous tumor from my pancreas. I had a very rare form of pancreatic cancer called an islet cell neuroendocrine tumor, which represents about 1 percent of the total cases of pancreatic cancer diagnosed each year, and can be cured by surgical removal if diagnosed in time (mine was). I will not require any chemotherapy or radiation treatments.
>
> —Steve Jobs[11]

2005—ALL THINGS DIGITAL, D5

In 2005, Steve Jobs and Bill Gates sat for an interview at D5 (The fifth All Things D) and made comments about their firms and each other. Jobs talked about how Apple really needed help with their programming language BASIC, which they received from Microsoft for $31,000. Steve Jobs and Bill Gates commiserated about starting the firm with young people and thirty years later being the oldest people at their organizations. And despite being seen in the popular media, as adversaries the two friends were willing to publicly praise the work of each other's firms.

Steve Jobs commented on the belief that festered in Apple that the competitor was really Microsoft, when that was never the intent.

> But the net result of it was, was there were too many people at Apple and in the Apple ecosystem playing the game of, for Apple to win, Microsoft has to lose. And it was clear that you didn't have to play that game because Apple wasn't going to beat Microsoft. Apple didn't have to beat Microsoft. Apple had to remember who Apple was because they'd forgotten who Apple was.
>
> —Steve Jobs[12]

While Apple was consistently known for the introduction of products, Jobs brought over the products and lessons of NeXT. He now internally viewed Apple as a software company, not a hardware company.

> And so the big secret about Apple, of course–not-so-big secret maybe–is that Apple views itself as a software company and there aren't very many software companies left, and Microsoft is a software company. And so, you know, we look at what they do and we think some of it's really great, and we think a little bit of it's competitive and most of it's not. You know, we don't have a belief that the Mac is going to take over 80 percent of the PC market. You know, we're really happy when our market share goes up a point and we love that and we work real hard at it, but Apple's fundamentally a software company and there's not a lot of us left and Microsoft's one of them.
>
> —Steve Jobs[13]

One of the ways Jobs reshaped Apple from integrated hardware to software was by removing vestiges of the past, not only from the years he was at Apple but the 12 years he was away. The history of Apple was effectively history, and the history was sent to a museum to clear the way for Apple's future initiatives without reminders and clutter from the past.

> And, you know, one of the things I did when I got back to Apple 10 years ago was I gave the museum to Stanford and all the papers

> and all the old machines and kind of cleared out the cobwebs and said, let's stop looking backwards here. It's all about what happens tomorrow. Because you can't look back and say, well, gosh, you know, I wish I hadn't have gotten fired, I wish I was there, I wish this, I wish that. It doesn't matter. And so let's go invent tomorrow rather than worrying about what happened yesterday.
>
> —Steve Jobs[14]

While Apple was known for integration of products, software, and services, Jobs lamented that Apple had to focus on the skills Apple did best, not what Apple could do but others did better. When others were better at a task, Apple would seek partnerships and trade, as the law of competitive advantage would suggest. Thinking back through the three-decade history of Apple and Microsoft, he admitted that Apple probably would have been a more successful firm if they were willing to partner with other firms as freely as Microsoft had throughout the firms' history.

> We don't think one company can do everything. So you've got to partner with people that are really good at stuff ... You know, because Woz and I started the company based on doing the whole banana, we weren't so good at partnering with people. And, you know, actually, the funny thing is, Microsoft's one of the few companies we were able to partner with that actually worked for both companies. And we weren't so good at that, where Bill and Microsoft were really good at it because they didn't make the whole thing in the early days and they learned how to partner with people really well.
>
> —Steve Jobs[15]

2006—MICROSOFT ZUNE AND OPTIONS BACKDATING SCANDAL

Jobs opened up to Steven Levy as to why the iPod had been successful when competitors had not been. Jobs believed the primary factor was the use of iTunes software to externally manage music, rather than making the hardware itself more complex; he had made the comment with Bill Gates that Apple was a software company. Jobs also opened up on why he felt that Microsoft's music player was doomed to failure.

> *Levy*: Other companies had already tried to make a hard disk drive music player. Why did Apple get it right?
> *Jobs*: We had the hardware expertise, the industrial design expertise and the software expertise, including iTunes. One of the biggest insights we have was that we decided not to try to manage your music library on the iPod, but to manage it in iTunes. Other companies tried to do everything on the device itself and made it so complicated that it was useless.
> *Levy*: Microsoft has announced its new iPod competitor, Zune. It says that this device is all about building communities. Are you worried?
> *Jobs*: In a word, no. I've seen the demonstrations on the Internet about how you can find another person using a Zune and give them a song they can play three times. It takes forever. By the time you've gone through all that, the girl's got up and left! You're much better off to take one of your earbuds out and put it in her ear. Then you're connected with about two feet of headphone cable.[16]

In 2001, Bill Gates had not been impressed by the Apple iPod, stating there was nothing that Microsoft could not do and that consumers would want choice. Microsoft's Zune product was released five years later as a competitor and discontinued in 2012.

OPTIONS BACKDATING SCANDAL

As 2006 closed, Apple faced a backdating scandal on employee options. Employee stock options give the employee the right (but not the requirement) of purchasing shares of stock at a later point in time. Options to purchase shares have two major components—a strike price and exercise dates. The strike price is what an employee would have to pay to acquire a share. The exercise dates are the timeframe when the employee can actually buy the shares under the option grant.

Options are priced on the day of the option grant. An employee might be granted an option today to purchase 10,000 shares at $20 each at some point three to five years from now, based upon the current day's price (say $20). But if the share price today is $20 and was $15 three months ago, an option grant "backdated" to three months ago would give the employee the future right to purchase shares at $15. And that's a form of fraud.

Apple conducted an independent review, undertaken by independent members of Apple's Board of Directors (directors who were not employees). One of the committee members was former Vice President Al Gore. The committee's findings were released to the media in October and submitted to the government in December.

> The internal review and the Special Committee's independent investigation identified a number of occasions between October 1996 and January 2003 (the "relevant period") when the Company used incorrect measurement dates for stock option grants. The independent investigation also found that during the relevant period:
>
> - Procedures for granting, accounting, and reporting of stock option grants did not include sufficient safeguards to prevent manipulation
> - The grant dates for a number of grants were intentionally selected in order to obtain favorable exercise prices
> - Two former officers of the Corporation engaged in conduct that raises serious concerns in connection with the granting, accounting, recording, and reporting of stock options
> - CEO Steve Jobs was aware or recommended the selection of some favorable grant dates, but he did not receive or financially benefit from these grants or appreciate the accounting implications
>
> —Special Committee of Apple's Board of Directors[17]

The Committee mentions Steve Jobs by name, and that he was involved in the backdating in some way, perhaps even recommending the backdating. In the initial press release to the public from Apple (before the submission of the document to the Security and Exchange Commission) in October 2006, the language differed.[18] There was no reference that Jobs had recommended the selection of some favorable grant dates. The public was told in October that Jobs was "aware," while the government was told in December that he was "aware or recommended."

2007—THE YEAR OF IPHONE

The minute Steve Jobs showed the original iPhone to the world, suddenly people didn't want their [expletive] flip phones anymore.

—Apple employee Matt MacInnis, founder of Inkling[19]

One time Steve said, "You know, everybody has a cell phone, but I don't know one person who likes their cell phone. I want to make a phone that people love." That was the foundation of what became the iPhone.

—John Lasseter[20]

BILL GATES (INADVERTENTLY) INSPIRING THE IPHONE

The next major Apple release of the first decade of the 2000s was the iPhone. A Microsoft engineer inadvertently—and indirectly—launched that idea. The engineer, who developed the tablet PC, was married to a friend of Steve Jobs. The engineer kept talking about the tablet PC he developed for Microsoft at a birthday party that was attended by both Jobs and Bill Gates. Neither party was happy, according to Isaacson, but for different reasons. Gates did not like disclosing intellectual property.[21]

He's our employee and he's revealing our intellectual property.

—Bill Gates[22]

Jobs was unhappy hearing about the Microsoft tablet project and the insistence from Gates's staff that using a stylus as pen was the direction of the future for tablets, so he ordered his staff to create a touch screen. Apple then realized the phone might have more immediate appeal than the tablet. Apple really created the touch-screen interface that was first used for the iPhone, then iPad, because Steve Jobs was upset after listening to a Microsoft engineer.[23]

INTRODUCTION OF THE IPHONE

On January 8, 2007, Steve Jobs spoke at MacWorld 2007. For Apple, this was a momentous day, as it was the introduction of the iPhone. Jobs hearkened back to 1984 and 2001, where Macintosh and iPod were introduced. One could question whether the Apple/Apple II (in 1976 and 1977) should be cited instead, but Macintosh, iPod, and iPhone were steps in a carefully constructed narrative.

> This is a day I've been looking forward to for two-and-a-half years ... Every once in a while a revolutionary product comes along that changes everything. One is very fortunate if you get to work on just one of these in your career. Apple has been very fortunate that it's been able to introduce a few of these into the world. In 1984 we introduced the Macintosh. It didn't just change Apple, it changed the whole industry. In 2001 we introduced the first iPod, and it didn't just change the way we all listened to music, it changed the entire music industry.
>
> —Steve Jobs[24]

Jobs then spoke about the iPhone, removing keyboards by making touch keys available on the screen when needed, and revolutionizing how touch screen works. And rather than using the stylus preferred by Bill Gates and the engineer at Microsoft (to which Jobs exclaimed "Yuck!"), Apple introduced Multi-Touch for the first time.[25] And in a not-so-subtle warning to other manufacturers who would want to use the innovation, Jobs stated "boy have we patented it."[26]

So what is Multi-Touch? Any touch screen where you can use multiple fingers at a time—for instance, to zoom in/out or highlight text—uses Apple's Multi-Touch patents. The connection between Macintosh, iPod, and iPhone were that each introduced a new way of interacting with a device—mouse, click-wheel, and Multi-Touch. Jobs was that careful in the construction of his speeches.

With iPhone, Apple focused on extensive intellectual property, and Jobs reiterated that fact to any and all of Apple's competitors. "We filed for over 200 patents for all the inventions in iPhone and we intend to protect them."[27]

Like most hardware releases from Apple under Steve Jobs, an official from Microsoft issued an insult about the new release:

> There's no chance that the iPhone is going to get any significant market share. No chance. It's a $500 subsidized item. They may make a lot of money. But if you actually take a look at the 1.3 billion phones that get sold, I'd prefer to have our software in 60 percent or 70 percent or 80 percent of them, than I would to have 2 percent or 3 percent, which is what Apple might get.
>
> —Steve Ballmer, Microsoft CEO[28]

And the stylus to which Steve Jobs claimed "Yuck!"? A few years after his death, Apple released the Apple Pencil, a stylus when users need to be more precise on the iPad than using fingers and Multi-Touch. Like Apple Music, this was another example of Apple developing or marketing a product that expressly did not meet Jobs's approval within a few years of his death.

GETTING TO THE INTRODUCTION

While the unveiling of the iPhone at MacWorld introduced a revolutionary product that would be available two quarters later, there's a story in the background that wasn't released until much later. The prototype iPhones demonstrated at the announcement really didn't work—Jobs had to follow a precise script and order of operations to get through the detailed demonstration. Even then, "the iPhone was still randomly dropping calls, losing its Internet connection, freezing or simply shutting down" on the day before the demonstration.[29] Obviously, Jobs would have a problem if these issues occurred during the initial public demonstration.

> At first it was just really cool to be at rehearsals at all—kind of like a cred badge … But it quickly got really uncomfortable. Very rarely did I see him become completely unglued—it happened, but mostly he just looked at you and very directly said in a very loud and stern voice, "You are [expletive] up my company," or, "If we fail, it will be because of you." He was just very intense. And you would always feel an inch tall.
>
> —Andy Grignon[30]

Jobs was characteristically intense and consumed with the idea that the iPhone demonstration had to work flawlessly. Compounding the issue, he was also insistent that the demonstration had to be a good experience for the audience. That meant adding more technology and cables to a prototype phone that was already malfunctioning.

> Jobs wanted the demo phones he would use onstage to have their screens mirrored on the big screen behind him. To show a gadget on a big screen, most companies just point a video camera at it, but that was unacceptable to Jobs. The audience would see his finger on the iPhone screen, which would mar the look of his presentation. So he had Apple engineers spend weeks fitting extra circuit boards and video cables onto the backs of the iPhones he would have onstage. The video cables were then connected to the projector, so that when Jobs touched the iPhone's calendar app icon, for example, his finger wouldn't appear, but the image on the big screen would respond to his finger's commands. The effect was magical.[31]

In addition, to get the WiFi to work reliably, the antennas had wires running off the stage and the WiFi network was set up using a frequency not even allowed in the United States. The fear was that the attendees—sophisticated in technology—might be on the network and interfere with the demonstration. Jobs's desires created many potential failure points within the demonstration. And if any member of the team told anyone of their work on this revolutionary product, firing was the most likely outcome.[32]

JOBS DOES NOT KNOW CUSTOMERS' REACTIONS

After introducing the new iPod, Jobs finally admitted being nervous before presentations but also reveals a concept that had previously been missing when creating new products; Apple was actually listening to customer feedback and desires when creating products.

> When we create stuff, we do it because we listen to customers, get their inputs and also throw in what we'd like to see, too. We cook up new products. You never really know if people will love them

as much as you do. The most exciting thing is you have butterflies in your stomach in the days leading up to these events. To learn people love it as much as you do is a relief and also really exciting.

—Steve Jobs[33]

2008—MAKING EMPLOYEES THE BEST POSSIBLE

Jobs's personal abuses are also legend: He parks his Mercedes in handicapped spaces, periodically reduces subordinates to tears, and fires employees in angry tantrums. Yet many of his top deputies at Apple have worked with him for years, and even some of those who have departed say that although it's often brutal and Jobs hogs the credit, they've never done better work.

—*Peter Elkind*[34]

On his demanding reputation, Jobs's response was simple and succinct:

My job is to not be easy on people. My job is to make them better.

—Steve Jobs[35]

When I hire somebody really senior, competence is the ante. They have to be really smart. But the real issue for me is, are they going to fall in love with Apple? Because if they fall in love with Apple, everything else will take care of itself. They'll want to do what's best for Apple, not what's best for them, what's best for Steve, or anybody else.

—Steve Jobs[36]

Jobs spoke about training every individual member of his executive team to one day follow him; every member of executive team was expected to be capable of serving as the next CEO.

We've got really capable people at Apple. I made Tim [Cook] COO and gave him the Mac division and he's done brilliantly. I mean, some people say, "Oh, God, if [Jobs] got run over by a bus, Apple would be in trouble." And, you know, I think it wouldn't be a party, but there are really capable people at Apple. And the board would have some good choices about who to pick as CEO. My job is to make the whole executive team good enough to be successors, so that's what I try to do.

—Steve Jobs[37]

APPLE CARS

Ever an innovator, Steve Jobs even had early discussions about what would constitute an Apple car with colleague Tony Fadell; given the financial crisis of the timeframe, Jobs decided 2008 would not be a good time to initiate work on an Apple car.

> "We had a couple of walks," Fadell told Bloomberg. The pair posed hypothetical questions to each other, such as: "If we were to build a car, what would we build? What would a dashboard be? And what would this be? What would seats be? How would you fuel it or power it?"[38]

ADDITIONAL CONCERNS WITH JOBS'S HEALTH

In July 2008, there were more worries about Jobs's health, when Apple stated that Jobs was afflicted with a common bug. Joe Nocera of *The New York Times* was one of the reporters who was unaccepting of that answer, and he pressed his luck with Jobs. After multiple inquiries, Nocera received a phone call.

> This is Steve Jobs … You think I'm an arrogant [expletive] who thinks he's above the law, and I think you're a slime bucket who gets most of his facts wrong.
>
> —Steve Jobs[39]

Jobs did agree to talk with Nocera off the record about his medical condition, refusing to allow the conversation to be released. Nocera reported at the time that "while his health problems amounted to a good deal more than 'a common bug,' they weren't life-threatening and he doesn't have a recurrence of cancer."[40]

2009—MAJOR SURGERY FOR JOBS

In his last years of life, Jobs allowed Walter Isaacson into his life to write a sole authorized biography, which was rushed into press after his death. Isaacson had clear instructions from Laurene Powell-Jobs.

> Be honest with his failings as well as his strengths. There are parts of his life and personality that are extremely messy. You shouldn't whitewash it. I'd like to see that it's all told truthfully.
>
> —Laurene Powell-Jobs[41]

In terms of his messy and truthful life, Isaacson simply declared:

> He's not warm and fuzzy.
>
> —Walter Isaacson[42]

CONTINUING HEALTH CONCERNS

The health concerns with Steve Jobs continued into 2009. In early January, Jobs sent a communication to Apple employees that his doctors had recently discovered a hormone imbalance.

> Fortunately, after further testing, my doctors think they have found the cause—a hormone imbalance that has been "robbing" me of the proteins my body needs to be healthy. Sophisticated blood tests have confirmed this diagnosis.
>
> —Steve Jobs[43]

Just a few months later, he flew to Memphis for a liver transplant.[44]

While Steve Jobs required a liver transplant to extend his life; he really didn't have to wait for the liver to become available in Tennessee because he already had a match proven via medical testing. Not a family member and not a stranger, but the man who would succeed him as CEO of Apple, then-COO Tim Cook. Cook had undergone testing that demonstrated he would have been an acceptable donor for a partial liver; Cook was not expected to have any unforeseen side effects and the removed part of the liver would be expected to re-grow. Not only did Jobs refuse the offer, Cook relayed that it was one of (comparatively few) times Jobs yelled at him.

> Somebody that's selfish ... doesn't reply like that. I mean, here's a guy, he's dying, he's very close to death because of his liver issue,

> and here's someone healthy offering a way out. I said, 'Steve, I'm perfectly healthy, I've been checked out. Here's the medical report. I can do this and I'm not putting myself at risk, I'll be fine.' And he doesn't think about it. It was not, 'Are you sure you want to do this?' It was not, 'I'll think about it.' It was not, 'Oh, the condition I'm in ... ' It was, 'No, I'm not doing that!' He kind of popped up in bed and said that. And this was during a time when things were just terrible. Steve only yelled at me four or five times during the 13 years I knew him, and this was one of them.
>
> —Tim Cook[45]

Years later, more details became known of the process by which Jobs had received a transplant in Memphis, Tennessee, flying in his private jet from California overnight when he was granted a liver from a young man who died in a car accident. After Jobs's surgery, his transplant surgeon lived in Jobs's Memphis house over the following two years, either on a part-time or full-time basis. Jobs's lawyer paid for the taxes and utilities on this house during that time, then sold the house to the surgeon for the same cost as Jobs's initial purchase. The actions of allowing his surgeon to live in the house—and covering some expenses—created the appearance that he had received preferential treatment for a liver transplant.[46]

2010—SCULLEY TRIES TO MEND FENCES

With years to reflect, John Sculley eventually began to speak about Steve Jobs again. Remarkably, he agreed with Steve Jobs's 1996 assessment that he had hired the wrong guy.

> Looking back, it was a big mistake that I was ever hired as CEO. I was not the first choice that Steve wanted to be the CEO. He was the first choice, but the board wasn't prepared to make him CEO when he was 25, 26 years old ... The reason why I said it was a mistake to have hired me as CEO was Steve always wanted to be CEO. It would have been much more honest if the board had said, "Let's figure out a way for him to be CEO. You could focus on the stuff that you bring and he focuses on the stuff he brings."
>
> —John Sculley[47]

In 1993, instead of looking for a firm that was willing to buy Apple, Sculley should have asked Jobs back (although Jobs was with NeXT and Pixar at the time):

> But if I had any sense, I would have said: "Why don't we go back to the guy who created the whole thing and understands it. Why don't we go back and hire Steve to come back and run the company?"
>
> It's so obvious looking back now that that would have been the right thing to do. We didn't do it, so I blame myself for that one. It would have saved Apple this near-death experience they had.
>
> —John Sculley[48]

Sculley believed that in the late 1990s Jobs himself was the only hope of saving Apple

> I'm actually convinced that if Steve hadn't come back when he did if they had waited another six months—Apple would have been history. It would have been gone, absolutely gone. What did he do? He turned it right back to where it was—as though he never left. He went all the way back.
>
> —John Sculley[49]

Sculley also chances into a discussion where he talks about a run-in where Apple did not have high enough standards, and it cost Apple rejected shipments in Japan. Attention to detail at the obsessive level of Steve Jobs really was the difference between failure (with lawsuits) and success in Japan.

> A big part of it was that we had to learn to make products the way the Japanese wanted products. We were assembling products in Singapore and sending them to Japan. And the first thing the customer saw when they opened the box was the manual, but the manual was turned the wrong way around—and the whole batch was rejected. In the United States, we'd never experienced anything like that. If you put the manual in this way or that way—what difference did it make?
>
> —John Sculley[50]

But Sculley also understood there were hard feelings with Jobs that were irreparable from Jobs's perspective.

> Q: People say he killed the Newton—your pet project—out of revenge. Do you think he did it for revenge?
> *Sculley*: Probably. He won't talk to me, so I don't know.[51]

IPHONE 4 RECEPTION ISSUES

Upon release, early iPhone 4 devices had a problem with reception that needed to be resolved. There were widespread complaints from customers that holding the iPhone in a certain manner led to poor reception and dropped calls. As a CEO that occasionally wrote responses to customer complaints, Jobs's advice to customers was succinct, attributing much of the blame for the issue to his customers:

> Just avoid holding it in that way.
>
> —Steve Jobs[52]

This was just one of many interactions directly with customers who communicated with him. In fact, his responses were frequent enough to result in compilations online and in books. Any response from Jobs might range from a single word, such as "no" to a few sentences. He also refused to allow customers to twist the truth or seek repairs of Apple products for reasons not covered under warranty. He was contacted by an individual with a water-damaged laptop complaining about a lack of a repair under warranty; the customer was harshly rebuked as simply deflecting blame:

> This is what happens when your MacBook Pro sustains water damage. They are pro machines and they don't like water. It sounds like you're just looking for someone to get mad at other than yourself.
>
> —Steve Jobs[53]

And while Jobs's marketing skill was clear in product releases, he was not concerned about displaying that skill in e-mails.

> This is not exactly high level stuff, and Jobs's replies aren't those that would have been crafted by some marketing exec. But there is no doubt that the information contained is useful, and it is widely reported when it occurs. Not to mention the fact that the recipients are no doubt very pleased to have had a response at all.[54]

After Jobs released the iPad in 2010—the last major release of Jobs time at Apple—his friend and competitor Bill Gates was still stuck on the idea of a stylus. And a keyboard.

> You know, I'm a big believer in touch and digital reading, but I still think that some mixture of voice, the pen and a real keyboard—in other words a netbook—will be the mainstream on that," he said. "So, it's not like I sit there and feel the same way I did with iPhone where I say, 'Oh my God, Microsoft didn't aim high enough.' It's a nice reader, but there's nothing on the iPad I look at and say, 'Oh, I wish Microsoft had done it.'"
>
> —Bill Gates[55]

2011—THE END OF THE JOBS IERA AT APPLE

In early 2011, a scholarly journal article was published asking if investors in Apple truly cared about the health of Jobs. If investors truly believed Jobs was indispensable to Apple, the stock price would be expected to fall drastically on any news of the CEO's poor health. Studying the effects of previous health announcements on the stock of Apple, the researchers discovered that rather than investors being worried about Jobs being indispensable to Apple, the media and reporters might be creating the story that Jobs was indispensable to the firm.

> The "but for" impacts of adverse information about Jobs's health are not huge in relative terms and range between zero and four percent of Apple's share price. These magnitudes belie suggestions in the trade press that Jobs's value to Apple was $20 billion or greater. If investors truly believe that Jobs is crucial to Apple's future, then they are rather timid, and even confused, in demonstrating that

conviction. At the end of the day, it appears that the media gives more credence to Jobs being irreplaceable than investors.[56]

The authors of the study had a hypothesis that would be readily tested. Jobs would leave Apple for good just four short months later.

RESIGNATION

On August 24, 2011, Jobs resigned as CEO of Apple, elevating himself to the position of Chairman of the Board:

> I have always said if there ever came a day when I could no longer meet my duties and expectations as Apple's CEO, I would be the first to let you know. Unfortunately, that day has come.
>
> —Steve Jobs[57]

This was Jobs's second resignation letter from Apple. This time, Jobs would not have an opportunity to come back to the firm. Yet at the time, Jobs's absence was expected to be temporary; the story was not expected to end yet. Jobs's replacement as CEO would later relay:

> Every time I saw him he seemed to be getting better. He felt that way as well. Unfortunately, it didn't work out that way.
>
> —Tim Cook[58]

Part 5

JOBS OUTSIDE WORK

Chapter 11

FAMILY AND RELIGION

Steve Jobs spoke glowingly of his "parents"; his adopted parents Paul and Clara Jobs were his parents. After the death of Clara, he would seek out his biological mother, discover he had a biological sister, and learn that he had managed to meet his biological father out of pure coincidence.

BIOLOGICAL PARENTS—ABDULFATTAH JANDALI AND JOANNE SCHIEBLE

As an adoptee, Jobs always knew that he was *lucky* to be alive, although he did have doubts at some points in time as to whether he was simply abandoned and unwanted by his biological parents. As an adult, Jobs was exceptionally grateful for his mother choosing to give birth rather than seek an abortion.

In his early 30s, he sought out his birth mother, discovering he also had a sister born of she and his father. Jobs only reached out for his biological mother after Clara Jobs died, with the permission of Paul Jobs.[1] His gratitude was solely to his biological mother; he never sought out his biological father, even ignoring outreach. Desmond wrote of Jobs's decision:

> Decades later, Jobs would express gratitude that his mother didn't abort him. And his own experience confirms that an unwanted child can ultimately reverse the pattern of male irresponsibility bequeathed to him by his biological father.
>
> Isaacson traces Jobs's effort to find his biological mother, a Midwestern graduate student raised in a Catholic family. "I wanted to meet my biological mother mostly to see if she was okay and to thank her, because I'm glad I didn't end up as an abortion. She was 23 and went through a lot to have me," Jobs told his biographer.[2]

Joanne Schieble had gone through a lot to give birth to Steve. She returned to the United States and the Jobs family was her second choice for adopting her child. A condition of the adoption was that Paul and Clara send the child to college. Jobs did go to Reed College; he just did not graduate. And when Jobs met his biological mother, he would learn he had a biological sister, with the same mother and father, and she was a novelist.

Sister Mona Simpson was able to articulate why her grandparents (also Jobs's grandparents) had been so opposed to their parents, Abdulfattah Jandali and Joanne Schieble, marrying in the mid-1950s.

> They weren't happy. It wasn't that he was Middle Eastern so much as that he was a Muslim. But there are a lot of Arabs in Michigan and Wisconsin. So it's not that unusual. My mom met my dad at the University of Wisconsin. He was her teacher. They were the same age but he'd gotten his PhD really young.
>
> —Mona Simpson, on her parents' meeting[3]

Jobs was born while Schieble's father was still alive; Schieble and Jandali married after her father passed away, then divorced when his sister Mona was four. Their father, Jandali, effectively disappeared from his sister's life as well. Joanne Schieble (now Joanne Jandali) remarried, taking the name Simpson—sister Mona Jandali became Mona Simpson.

Jobs maintained his relationship with his biological mother for the rest of his life. In addition to attending events with his biological mother and sister, he also noted that he "does keep in touch with Joanne Simpson and invites her to some of his family gatherings."[4]

ABDULFATTAH JANDALI

Jobs never reached out to this biological father, although Jandali noted on multiple occasions he would have enjoyed speaking to his son once. Jandali would state that he did not become aware that his son was Steve Jobs until approximately 2005, when his son was already 50.

In the last years of Jobs's life, Jandali reached out to his son via e-mail on birthdays, and hoping for a positive result in the battle against cancer. While there's a dispute as to whether Jobs ever acknowledged his father, any response was (at most) two words.

> Jobs was, at best, courteously curt when Jandali wrote him happy birthday sentiments, or to wish him well in his battle with cancer. According to Jandali, Jobs responded with a one-sentence email: "Thank you." Sources close to the Jobs family tell the *WSJ* [*Wall Street Journal*] that Jobs didn't respond at all.[5]

After serving as a college professor, Jandali had gone into casino operations; at the time of Jobs's death, he was the general manager of a casino outside Reno, Nevada. He was described as using Apple products after learning his son was Steve Jobs (around 2005). Of his wife Joanne and daughter Mona, "[h]e abandoned the family" and was "for the most part unreachable." Like his son, he was described as knowing what customers wanted, although he completed diverged from his son by rarely taking the spotlight:

> He is really the opposite of a showman because he would always put the light on others to take the stage. He understands what guests like and what they are willing to pay for.[6]

In the days after Jobs resigned as CEO of Apple, Jandali expressed hopes that he would get to speak to his son. He described being an absentee father as a haunting experience but that the only way he would speak to his son would be for Jobs to call.

> This might sound strange, though, but I am not prepared, even if either of us was on our deathbeds, to pick up the phone to call him … Steve will have to do that, as the Syrian pride in me does not want him ever to think I am after his fortune.
>
> —Abdulfattah Jandali[7]

> I really am not his dad. Mr. and Mrs. Jobs are, as they raised him. And I don't want to take their place.
>
> —Abdulfattah Jandali[8]

There was one interesting circumstance that came out late in October 2011, the month Jobs died, when Walter Isaacson's biography of Jobs was released. Jobs and Jandali had met decades before and Jobs had known he had met his biological father since the mid-1980s; Jobs learned about the connection after his sister Mona Simpson had a meeting with their biological father.

> [Jandali] told her, somewhat emotionally, that he wished she could have seen him when he was managing a Mediterranean restaurant north of San Jose. "That was a wonderful place," he said. "All of the successful technology people used to come there. *Even Steve Jobs.*" Simpson was stunned. "Oh, yeah, he used to come in, and he was a sweet guy, and a big tipper," her father added. Mona was able to refrain from blurting out, *Steve Jobs is your son!* … Jobs was understandably astonished when she mentioned the restaurant near San Jose. He could recall being there and even meeting the man who was his biological father. "It was amazing," he later said of the revelation. "I had been to that restaurant a few times, and I remember meeting the owner. He was Syrian. Balding. We shook hands."[9]

Jobs had indeed met his biological father by happenstance, learned he was his biological father later, and had no interest in making any form of familiar connection or recognition.

BIOLOGICAL SISTER—MONA SIMPSON

For Joanne,
our mother,
and
my brother Steve

—Dedication of Mona Simpson's *Anywhere but Here*, in 1986[10]

In 1986, Steve Jobs is 31 and Mona Simpson acknowledges her brother in the dedication to her book *Anywhere but Here*, using only his first name. For the release of this book, George Plimpton of *The Paris Review* threw a party. Mona arrived not only with her mother but her brother, Steve Jobs.

> I had known Mona for quite a while ... She had said she had a brother who worked in the computer industry. But that party was the first time I learned that her brother was Steve Jobs.
>
> —Amanda Urban, Simpson's literary agent[11]

Given Jobs's penchant for privacy, this wouldn't be a surprise even if the two had always known each other. Simpson and Jobs had just recently met; Jobs didn't seek out his mother (and then sister) until Clara Jobs had died.

The next year, she talked about her family in a magazine, noting that her grandparents had not accepted her mother and father as a couple. She does not refer to her brother by name but suggests that she should learn some Arabic.

> My cousins and aunts and uncles still live in small cities in Wisconsin and Michigan and Illinois. That's my mother's family. My father is Syrian and most of his family lives in a village in Syria called Homs ... I should learn Arabic. It's terrible. My brother doesn't know it either. None of us do.
>
> —Mona Simpson[12]

After Simpson's *Anywhere but Here*, she wrote two more books related to Steve Jobs and the dynamics of their unique family. The first is about their father, who had been absent for both of them. The second was inspired directly by her brother. As would be later described in *Esquire*, the first book was:

> *The Lost Father*, out of their common haunting, out of her dreams of the man who, to some degree, had left them both. Then she wrote *A Regular Guy*. It is, in fact, about a singular guy. It is about a millionaire entrepreneur who, from the first page, carries around "this inability, not just to pander, but to see any need to pander to the wishes and whims of other people" and wants "the face of the earth to look different after his life upon it."[13]

The abandonment by her father at the age of four bothered her; she always made assumptions that the primary man in her life would be her father. Instead, it was the brother she had never known about until adulthood.

> Even as a feminist, my whole life I'd been waiting for a man to love, who could love me. For decades, I'd thought that man would be my father ... I met that man and he was my brother.
>
> —Mona Simpson[14]

Mona Simpson would marry a television writer (Richard Appel), who was involved with the cartoon *The Simpsons*. Even among those who have not read her novels, she may also be known through a separate popular culture reference. Husband Appel named Homer Simpson's cartoon mother "Mona Simpson" after his wife; he realized there was never an episode of the cartoon addressing the name of Homer Simpson's mother.[15]

In her 2011 eulogy for her brother, Simpson spoke of the deep connection that she had with her brother. The timeline of their meeting was a bit off; she described meeting her brother at 25 but was either 28 or 29 (after the death of Clara Jobs but before *Anywhere but Here* was released). Jobs had convinced her not to purchase a Cromemco computer, as "he was making something that was going to be insanely beautiful." She also revealed that her exceptionally wealthy brother always made a point to meet her at the airport, and that he may have been a mathematician if his childhood had been a little different.[16]

RELIGION

After renouncing religion after his Lutheran church experience in 1968, Jobs ascribed to Buddhist religious philosophies from his trip to India in 1974 through the rest of his life. Yet even his friend Wozniak spoke of never knowing Jobs to regularly attend any form of church services.

> Steve Jobs may be an informal fan of Eastern religions but it's never obvious in him and I never heard of him regularly attending a church.
>
> —Steve Wozniak[17]

But Jobs was dedicated to Zen Buddhism for a period of more than 35 years, and was considered to have many of the characteristics of a good Zen Buddhist.

> As a young seeker in the '70s, Jobs didn't just dabble in Zen, appropriating its elliptical aesthetic as a kind of exotic cologne. He turns out to have been a serious, diligent practitioner who undertook lengthy meditation retreats at Tassajara—the first Zen monastery in America, located at the end of a twisting dirt road in the mountains above Carmel—spending weeks on end "facing the wall," as Zen students say, to observe the activity of his own mind.[18]
>
> Like a Zen fussbudget, Jobs paid precise, meticulous, uncompromising attention to every aspect of the user experience of Apple's products—from the design of the fonts and icons in the operating system, to the metals used to cast the cases, to the colors on the boxes and in the magazine ads, to the rhyming proportions in the layout of Apple stores. He encouraged mindfulness in his customers too, by designing his computers so superbly that they faded into the background as creative imagination took over.[19]

Schlender and Tetzeli, in *Becoming Steve Jobs*, wrote that Zen Buddhism was a precise explanation for Jobs requiring others (and himself) to meet his standard of perfection.

> Certain elements of Buddhism suited him so well that they would provide a philosophical underpinning for his career choices—as

> well as a basis for his aesthetic expectations. Among other things, Buddhism made him feel justified in constantly demanding nothing less than what he deemed to be "perfection" from others, from the products he would create, and from himself.[20]

Jobs at times exhibited behaviors inconsistent with Buddhist and Eastern traditions. Milian noted that he possessed a

> ... complicated belief system that extends well beyond the Buddhist teachings ... Karma is another principle of the religion, but it didn't appear to be a system Jobs lived by. If he feared karma coming back to bite him, the sentiment wasn't evident in his public statements about competitors and former colleagues, calling them "bozos" lacking taste. ... Jobs spurned most reporters' interview requests, misled them in statements he did give, refused to disclose details of his cancer to investors until undergoing an operation and became shrouded in a scandal involving backdating stock options.[21]

One ex-Apple employee questioned whether Jobs could truly be considered a good Buddhist.

> The question is not whether he's an a**hole. That's beside the point. The question is whether he can be an a**hole and a good Buddhist.
>
> —Former Apple employee[22]

BACHELOR PARTY

While Jobs admittedly used illegal drugs in the 1970s, he was not known to drink alcohol very often. One notable example arose from his bachelor party, when Steve Jobs didn't know how to drink tequila (and sipped a single shot for an entire evening).

> Jobs, [Avie] Tevanian and one other friend went to a bar, where they had a tequila shot each. "Steve looks at it and kind of sniffs it, and [is] like, 'what am I supposed to do with it?' So we demonstrated the proper technique, and we said, 'now you try Steve',

and he wouldn't do it. He just sat there the whole night, sipping his shot of tequila." Jobs later sent Tevanian an email saying he'd had a great time with his true friends.[23]

MARRIAGE TO LAURENE POWELL

Sister Mona Simpson relayed being called by her brother the day he met Laurene Powell while providing a guest lecture at Stanford.

> There's this beautiful woman and she's really smart and she has this dog and I'm going to marry her.
>
> —Steve Jobs, per his sister Mona Simpson[24]

Information about the personal life of the Jobs family was tightly controlled. Jobs's future wife Laurene is known as "Lo" within her inner circle. She "grew up in New Jersey and is the daughter of a Marine pilot and a teacher. She attended the Stanford Graduate School of Business, where she met Mr. Jobs. They married in 1991 and had three children."[25]

After the two were "married in a small Buddhist ceremony in Yosemite National Park in 1991, and lived in Woodside, California,"[26] there was no plan for a honeymoon, as

> Every day with her is a honeymoon.
>
> —Steve Jobs[27]

Once married, Jobs spent most of his evenings at home with his wife and children, not traveling or attending events. While he could be preoccupied, he did reserve time for his family.

> He didn't go to a lot of fancy dinners. He didn't go out very much. He didn't go to galas or get himself feted. And once he had children, he didn't travel much. He came home and sat at a nice long wooden table in the kitchen of the house in Palo Alto. And even though he was often distracted, and even though he could be a bit snippy at times with people around him, he was there at dinner every night with his kids and his great wife.
>
> —Walter Isaacson[28]

Laurene shared characteristics with her husband, donating to the Democratic party, maintaining friendships with Bill and Hillary Clinton, and keeping a connection with Apple CEO Tim Cook.[29]

PARENTING

Steve Jobs was a parent to four children; one born out-of-wedlock in the 1970s (Lisa Brennan-Jobs) with Chrisann Brennan as the mother and three born in wedlock with Laurene Powell Jobs. Steve Jobs evolved as a father; he denied paternity of his oldest child initially but repaired relationships such that she would take his name and spend part of her childhood with Jobs, part with Chrisann Brennan.

LISA BRENNAN-JOBS

Lisa wrote of her upbringing, first detailing (as a student at Harvard) how her aunt's (Mona Simpson) book *A Regular Guy* reflected her life. The "regular guy" was modeled on Steve Jobs, and the character "Jane" seems inspired by Lisa. The factual framework from her life made a narrative suitable for a novel.

> *A Regular Guy* is, indeed, a work of fiction. The truths—points of commonalty between Jane and me—are mixed with equal or greater parts invention. Still, Jane bears a strong resemblance to me. Like me, Jane was born out of wedlock, grew up with a single mother, moved 13 times, and began the slow process of getting to know her father later in life. The book is cluttered with my life's details parading as Jane's—the "dangly" earrings I wanted to wear in sixth grade, descriptions of my old houses, how I ran for class president in high school. Like Jane, I was born in Oregon. My mother is an artist, my father an entrepreneur. Just like Jane's.
>
> —Lisa Brennan-Jobs[30]

Later, Lisa would write an article in *Vogue* about her experiences in Italy, which was described as more fulfilling than her childhood in Palo Alto, California.

> Eccentricities were celebrated, and no one was isolated. There was tradition and camaraderie, and all of it seemed more fulfilling

> than what I'd had growing up in Palo Alto, California. Italy was where the soul went to find calm and love, and I wanted to hold the best of it in the palm of my hand.
>
> —Lisa Brennan-Jobs[31]

She detailed how life varied back home in California between the time she spent with her mother and the time spent with her father.

> In California, my mother had raised me mostly alone. We didn't have many things, but she is warm and we were happy. We moved a lot. We rented. My father was rich and renowned and later, as I got to know him, went on vacations with him, and then lived with him for a few years, I saw another, more glamorous world. The two sides didn't mix, and I missed one when I had the other.
>
> —Lisa Brennan-Jobs[32]

She also felt that her treatment while in Italy may have been influenced by the fact that her father was Steve Jobs, not just on her character alone.

> Years later I wondered why I had been welcomed into this society without the social pedigree that usually enables such border crossings. At the time I thought it was luck—but perhaps I was naive. Marco's family knew that my father had money, and I wonder now if they assumed that I would one day inherit some of it from him, or if they were reassured by his cachet. Perhaps more than I understood at the time his name bought me admission to this Italy, as it would have in a story by Henry James. Their feelings for me were genuine, I knew, but maybe without this they would not have accepted me.
>
> —Lisa Brennan-Jobs[33]

AN ORDINARY PARENT

Steve Jobs was a fairly normal (although billionaire) parent later in life, particularly after marrying Laurene Powell. He developed his children by incorporating them into decisions, refused to allow technology to take over their lives, kept drawings from his children in his office, and even bonded with other young executives over the joy of being a parent.

INCORPORATION INTO DECISIONS

Although Jobs was a billionaire in 1996 (after the IPO of Pixar), he had a multiple-week conversation with his family—each night at the dinner table—about purchasing a new washer and dryer. Of his children, only Lisa would have been old enough to meaningfully contribute. Those discussions included design principles. And the billionaire Jobs complained about washers and dryers that cost thousands of dollars.

> Design is not limited to fancy new gadgets. Our family just bought a new washing machine and dryer ... It turns out that the Americans make washers and dryers all wrong. The Europeans make them much better—but they take twice as long to do clothes! It turns out that they wash them with about a quarter as much water and your clothes end up with a lot less detergent on them. Most important, they don't trash your clothes. They use a lot less soap, a lot less water, but they come out much cleaner, much softer, and they last a lot longer. We spent some time in our family talking about what's the trade-off we want to make. We ended up talking a lot about design, but also about the values of our family. Did we care most about getting our wash done in an hour versus an hour-and-a-half? Or did we care most about our clothes feeling really soft and lasting longer? Did we care about using a quarter of the water? We spent about two weeks talking about this every night at the dinner table.
>
> —Steve Jobs[34]

LOW-TECH PARENT

Like many other technology executives, Jobs limited the ability of his children to use technology, including products that were made by his own company. As Nick Bilton wrote about trying to lighten a tense conversation with Jobs:

> "So, your kids must love the iPad?" I asked Mr. Jobs ... The company's first tablet was just hitting the shelves. "They haven't used it," he told me. "We limit how much technology our kids use at home."

"Every evening Steve made a point of having dinner at the big long table in their kitchen, discussing books and history and a variety of things," he said. "No one ever pulled out an iPad or computer. The kids did not seem addicted at all to devices."[35]

DRAWINGS IN HIS OFFICE

When Tim Cook was discussing keeping Jobs's office precisely as he left it, he mentioned taking Laurene and Eve to the office in 2014; Jobs had left some of his children's drawings on his white board.

> But I wanted to keep his office exactly like it was. I was in there with Laurene [Powell Jobs, Steve's wife] the other day because there are still drawings on the board from the kids. I took Eve [Steve's daughter] in there over the summer and she saw some things that she had drawn on his white board years earlier.
>
> —Tim Cook[36]

SAME LONG-TERM RESIDENCE

His children are still raised in the same, functional but not pretentious home Jobs had purchased more than twenty years previously.

> Ms. Powell Jobs, a food lover, lives with her children in the same unpretentious red brick home she and Mr. Jobs bought two decades ago, where they raise bees and send friends Christmas baskets with hand-labeled Mason jars of honey.[37]

The Jobs residence was known for having the scariest house in the neighborhood every Halloween, between the special effects and the items given out each year. His family now gives out chocolate candy.

> Screams reverberated around the yard. Some were sound effects; some were real screams from freaked-out trick-or-treaters.
>
> Ewww: Odwalla bars. Every year. Rumor had it that Jobs's wife was a health nut. I never ate them.
>
> These days, though, the family hands out Toblerone chocolate bars.
>
> —Margaret Kadifa[38]

CONNECTING THROUGH PARENTING EXPERIENCES

When Jobs was negotiating with Microsoft in 1997, he bonded with the CFO (Greg Maffei) over a discussion of being a parent, "the 37-year-old Maffei had a young son, and Jobs, who dotes on his own children, told him what it was like to teach a kid to read."[39]

After Jobs's death, the media revealed information that showed how his children were shaped by him and the opportunities he could make available.

> College student Reed Jobs decided to study oncology after seeing his father, Apple CEO Steve Jobs, battle cancer. Reed's younger sister Eve is a great horseback rider, and their sister Erin has her father's great sense for design. And the eldest, Lisa, who was estranged from her father when she was young, became very close with him in recent years.[40]

Chapter 12

POLITICAL INVOLVEMENT

I've never voted for a presidential candidate. I've never voted in my whole life.

—Steve Jobs[1]

For an individual who self-reported as never voting through the mid-1980s, Jobs was involved in political discussions. Later in life, he was involved with multiple sitting U.S. presidents through a prospective nomination to one presidential commission, donations to one political party, a friendship with a sitting U.S. president, and advocacy for policies that would help his (and similar) firms attract employees.

THOUGHTS ON EDUCATION

Jobs and Apple recognized as early as 1979 that students could easily go through school without ever seeing a computer. As a business development strategy in 1979, Jobs recognized the company could provide at least one Apple II to every school in the nation at comparatively little cost through a tax incentive program. The Apple II series was the most common computer found in K–12 schools from 1979 until the 1990s.

> We wanted to donate a computer to every school in America. It turns out that there are about a hundred thousand schools in

> America, about ten thousand high schools, about ninety thousand K–8. We couldn't afford that as a company. But we studied the law and it turned out that there was a law already on the books, a national law that said that if you donated a piece of scientific instrumentation or computer to a university for educational and research purposes you can take an extra tax deduction. That basically means you don't make any money, you lose some but you don't lose too much. You lose about 10 percent. We thought that if we could apply that law, enhance it a little bit to extend it down to K–8 and remove the research requirements so it was just educational, then we could give a hundred thousand computers away, one to each school in America and it would cost our company ten million dollars which was a lot of money to us at that time but it was less than a hundred million dollars if we didn't have that.
>
> —Steve Jobs[2]

Despite giving a computer to every school in the United States in 1979, Jobs was also very clear in his belief computers don't replace people—there are many lessons that could be first inspired, and then taught, without computers, as had occurred with his own fourth-grade teacher.

> As you've pointed out I've helped with more computers in more schools than anybody else in the world and I am absolutely convinced that is by no means the most important thing. The most important thing is a person. A person who incites your curiosity and feeds your curiosity; and machines cannot do that in the same way that people can. The elements of discovery are all around you. You don't need a computer. Here—why does that fall? You know why? Nobody in the entire world knows why that falls. We can describe it pretty accurately but no one knows why. I don't need a computer to get a kid interested in that, to spend a week playing with gravity and trying to understand that and come up with reasons why.
>
> —Steve Jobs[3]

Jobs refined his beliefs over time as he gathered additional information and life experiences. He originally believed students would benefit from the exposure to computers. He later came to believe that the problems in K–12 education weren't fixable at all with technology; he perceived the problem as unions among K–12 educators.

> I used to think that technology could help education. I've probably spearheaded giving away more computer equipment to schools than anybody else on the planet. But I've had to come to the inevitable conclusion that the problem is not one that technology can hope to solve. What's wrong with education cannot be fixed with technology.
>
> No amount of technology will make a dent. It's a political problem. The problems are sociopolitical. The problems are unions. You plot the growth of the NEA [National Education Association] and the dropping of SAT scores, and they're inversely proportional. The problems are unions in the schools. The problem is bureaucracy.
>
> —Steve Jobs[4]

Jobs became an ardent proponent for school vouchers and school choice. He believed in a system whereby public schools would improve based upon having to compete for their funding, an idea espoused by economist Milton Friedman more than 25 years previously. If a local school system was budgeting a specific amount per student for the next school year, each parent would receive that amount as a voucher and then be allowed to choose a school, public or private. He gave an example of his daughter attending one of the best schools in America, which was only slightly more expensive than his state's spending on an average public school. Any family willing to spend the difference between a voucher and the cost of the school could send their child to the same school, and parents with vouchers might encourage the development of new, innovative schools.

> Mr. Jobs believes a full voucher system could save the education system. He relates that at the private school his daughter attended ["one of the 100 best schools in America"] tuition was $5,500 a year. "This is a lot of money for most parents," he concedes, "but the teachers [there] were paid less than public schools teachers—so

> it's not about money at the teacher level." Since California pays $4,400 a year to teach public school children, he reasons that "while there are not many parents who could come up with $5,500 a year, there are many who could come up with $1,100 a year." As a result, "schools would be starting up right and left."[5]

One author saw the connection between the hippie Steve Jobs of the 1970s and the Steve Jobs who was promoting vouchers as a way to improve schools; the issue was freedom and meritocracy. Under a mandate of meritocracy, good schools would seek good teachers and parents would use vouchers to reward those schools creating positive student outcomes, while underperforming schools would fail. The freedom to fail was an attitude Jobs brought back to Apple upon his return.

> Here, I think, is where the Jobs who lived a random life and the Jobs who wanted to bust the teachers' unions find common ground: he believed in freedom, including the freedom to fail hard. He supported meritocracy and opposed central planning.
>
> —Paul Wells[6]

IN GOVERNMENT RECORDS

When seeking security clearances for access to government documents or political appointments within the U.S. government, individuals undergo detailed background checks. Jobs underwent at least three checks of this type:

- A Department of Defense background check for top secret clearance in 1988 (while at Pixar)
- An FBI background check for potential appointment to President George H.W. Bush's President's Export Council in 1991
- A further Department of Defense background check in 1994

These files consist of the interviews with Jobs, searches of other public records, and interviews with personnel who knew Jobs well. Due to the Privacy Act in the United States, the written records of the interviews have the names of the persons being quoted redacted, unless it was the Subject (Jobs) himself.

In the 1988 and 1994 Department of Defense files, the following details are learned about Jobs:

> Jobs listed himself as attending Reed from 1972 to 1974, although he was only officially enrolled for the Fall 1972 semester.
>
> "Subject lives a very simple life, which includes being a vegetarian and keeping his body free of alcohol or drugs."
>
> [Redacted] remarked Subject is "fair with his workers, but sometimes hard to get along with due to mood swings." When asked about Subject's mental stability, [redacted] stated Subject was almost a "manic depressive" at time. [Redacted] stated Subject would be "really up" one day, and "really down" on others.
>
> "Subject stated he has a daughter who was born out wedlock and felt the possibility is there for her to be kidnapped."[7]

Upon reapplying for a security clearance in 1994, Jobs had forgotten details from 1988, such as the level of his previous clearance (which was at the highest level, top secret).[8]

REPUBLICAN PRESIDENTIAL ADMINISTRATION

President George H. W. Bush (41st President of the United States)

In 1991, President George H. W. Bush contemplated appointing Steve Jobs to the President's Export Council (PEC). As a result, the Federal Bureau of Investigation (FBI) was required to undertake the second background check; this check also used information from the Department of Defense check as corroborating evidence. The PEC's role is to determine what is required by various industries for U.S. producers to export products and services; Jobs would later have a dispute with a U.S. president when he wanted to produce goods in the United States, rather than making the goods abroad and then importing those items (the exact opposite of the PEC's purpose). Per the U.S. Department of Commerce, the duties of the PEC are to:

- survey and evaluate the export expansion activities of the communities represented by its membership; identify and examine

specific problems that business, industrial, and agricultural practices may cause for export trade; examine the needs of business, industry, and agriculture to expand their efforts; and recommend specific solutions to these problems and needs
- act as liaison among the communities represented by the membership and may provide a forum for those communities on current and emerging problems and issues in the field of export expansion
- encourage the business, industrial, and agricultural communities to enter new foreign markets and to expand existing export programs
- provide advice on Federal plans and actions that affect export expansion policies that have an impact on those communities represented by the membership[9]

The application authorizing this investigation showed Jobs submitted the initial authorization for the background investigation on February 6, 1991. As a deceased individual, his Social Security Number (549-94-3295) was not redacted on public records.

In the interviews conducted by the FBI, there are numerous references to the intellectual brilliance of Steve Jobs. Many of the respondents simultaneously spoke of behaviors that would be considered manipulative. Others spoke about behaviors that might be frowned upon if released to the public, such as previous drug use. Under the Privacy Act, some information that is widely known—such as the name of his adopted father—is among the information that is redacted.

While discussing his drug use in the 1970s (which was condensed to 1970–1974), Jobs again stated that he had never been arrested for any offense, despite signing an affidavit of the 1975 arrest on his Department of Defense background check just two-and-a-half years previously.

> He had never been charged, arrested or convicted of any offenses.[10]

Of the "reality distortion field" and intentionally deceptive behaviors through the 1980s,

> Several individuals questioned Mr. Jobs's honesty stating that Mr. Jobs will twist the truth and distort reality in order to achieve his goals. They also commented that, in the past, Mr. Jobs was not

supportive of [redacted] [the mother of his child born out of wedlock] and their daughter; however recently has become more supportive.

[Redacted] explained that, when Mr. Jobs resigned from ACI [Apple Computer Incorporated], he took proprietary information and key technological personnel with him. That prompted the lawsuit, which has since been resolved. ... He characterized Mr. Jobs as a deceptive individual who is not completely forthright and honest. He states that Mr. Jobs will twist the truth and distort reality in order to achieve his goals.

He described the Appointee as an individual who was not totally forthright and honest and has a tendency to distort reality in order to achieve his goals. He offered a comparison in that he [redacted] has high ethical standards and does business with people using this standard, however, the Appointee will twist the truth in order to achieve whatever goal he has set for himself. He therefore considered the Appointee to be a deceptive person.

... a visionary and charismatic individual who was at the same time shallow and callous to people in his personal relationships. She described his personal life as being lacking due to his narcissism and shallowness.

... although the Appointee is basically an honest and trustworthy person, he is a complex individual and his moral character is suspect. He stated he is no longer friends with the Appointee and feels bitter and somewhat alienated, based upon having to work for him at Apple. He explained that he [redacted] did not receive any stock as a result, which would obviously have made him quite wealthy now. He stated the Appointee alienated a large number of people at Apple, as a result of his ambition.[11]

Others spoke of Jobs in very positive terms, although there are often disagreements with contemporaries when it came to business matters:

[Redacted] described appointee as an extremely intelligent individual, a true leader, who has made a difference in the computer industry and given the opportunity will make a positive contribution on the National scene. [Redacted] stated that appointee is not an individual who can be intimidated; however, fits in and

> can talk with anyone. [Redacted] further described appointee as an extraordinary person, who is an excellent business negotiator and recruited of talent. [Redacted] added that appointee is a demanding individual, expecting a lot from himself and other. [Redacted] felt that most people including himself, believe appointee to be an extraordinary person with a good reputation, but when it comes to business there are frequently disagreements and sometimes hard feelings. [Redacted] added that some people have a great respect for appointee and others dislike him.[12]

In travel that would be of note to U.S. security in the early 1990s, it was declared that Jobs had been to the Soviet Union and Japan, but none of his other international travel was declared. The identification of the Soviet Union was of clear interest at the time, as the Soviet Union was not dissolved (first informally and then formally) until later that year.

> He had extensive foreign travel and had been to Japan and the Soviet Union.[13]

Close friends were asked whether Jobs lived within his financial means, with one responding that not only did he live within his means, he lived almost like a monk for someone of his level of wealth.

> He advised that the Appointee lives within his means financially, however based upon his newfound religious beliefs, the Appointee lives more of a spartanlike and at times even monastic existence.[14]

When the FBI was speaking to former neighbors, Steve Jobs's protection of his privacy was clearly established. One of his neighbors didn't even know him.

> ... she did not know the Appointee as a neighbor, even though he had lived in the area but that her husband knew the Appointee through his computer business.[15]

The FBI investigative file does disclose one major item that was not well known about Jobs. While at Pixar, Steve Jobs held a top secret (highest-level) government security clearance, issued after the 1988

Department of Defense background check. That information wasn't disclosed within the released Department of Defense files from 1988.

> Top Secret clearance dated November 3, 1988, based on a Background Investigation by the Defense Investigative Service dated August 30, 1988. This clearance terminated July 31, 1989, and the employing agency is:
> PIXAR
> San Rafael, California[16]

DEMOCRATIC PRESIDENTIAL ADMINISTRATIONS

President Bill Clinton (42nd President of the United States)

Steve Jobs was friendly with Democratic president Bill Clinton—successor to Republican George H.W. Bush—and provided major donations to the Democratic Party. His wife has continued the friendship with the Clinton family. Jobs was implicated in a scandal after President Clinton's first term, when it was revealed that the president authorized individuals to stay in the White House's Lincoln Bedroom in exchange for donations to the Democratic Party.

> White House documents, some in the President's own handwriting, indicate the Lincoln bedroom was at least on the market to major political donors. Under increasing public pressure, the White House released the names of 958 visitors who slept at the White House during Clinton's first term. Most were family friends, but many were major political contributors, like computer magnate Steve Jobs, who gave $150,000.
>
> —Jim Miklaszewski[17]

When Bill Clinton's daughter Chelsea decided to attend Stanford (near Palo Alto, California), the president already had a place to stay—he had a standing invitation to stay at the Jobs residence:

> What he gave me was the opportunity to see her. He gave me time with my daughter.
>
> —President Bill Clinton[18]

Despite their friendship, Clinton did not know whether Jobs would have been an acceptable politician,

> ... because you have to be somewhat more inclusive and you don't have the same amount of authority without checks.
>
> —President Bill Clinton[19]

President Barack Obama (44th President of the United States)

Walter Isaacson wrote of interactions between Jobs and President Barack Obama. Jobs "stressed the need for more trained engineers and suggested that any foreign students who earned an engineering degree in the United States should be given a visa to stay in the country. Obama said that could be done only in the context of the "Dream Act."[20] Hearkening back to the statement of President Bill Clinton that suggested Jobs might not be the best politician, the inability to take an action that would improve business infuriated Jobs. Jobs saw the president as providing reasons of why his idea was not possible, and stressed that having 30,000 additional trained production engineers in the United States would permit many more production facilities (and thus employment opportunities) for Apple and other technology companies in the United States.

After the release of Isaacson's biography of Jobs, his interactions with Obama were criticized.

> He [Jobs] even instructed Obama that the United States had to behave more like China in the manner in which it encouraged corporations to act free of regulations or concern for their employees and their environment.
>
> —Eric Alterman[21]

Chapter 13

THE MAN OUTSIDE APPLE

CONTINUING RELATIONSHIP WITH WOZNIAK

Although Jobs only lived 56 years, he knew Wozniak for more than 40 of them. Steve was in middle school and Wozniak in high school when they originally met; they had a friendship that continued from that point forward. Wozniak also admitted readily that Jobs could be a challenging and confrontational character but denied that Jobs ever treated him with those types of behaviors. He also wrote at times of ways in which he hoped his friendship could be improved and of some of the friction points between them.

> Many times I wish that we were close. Steve can relax and enjoy my many stories, whereas a lot of business driven people can't. He is more trapped to his job responsibilities and partly wishes that he could be like myself, with freedom and time for experiences with students and family.
>
> —Steve Wozniak[1]

As noted earlier, Wozniak solely wanted to be an engineer and programmer. He had no desire to be involved in the business aspect, showing his consistent competence in design and engineering. He did chafe of Jobs focusing on products such as the Apple III and Macintosh at times but noted "I've never seen Steve pursue less than the best and products that change things for the better."[2]

Wozniak also describes never actually seeing negative behavior from Jobs—Wozniak could always go to the engineering department for refuge. Yet he attests that Jobs had indeed matured between leaving Apple in 1985 and returning in 1997, stating:

> He seems to have mellowed and be more understanding, but I always saw him that way. I never saw his drastic actions.
>
> —Steve Wozniak[3]

While Wozniak stated he never saw poor behavior from Jobs, he also didn't deny that the negative behavior existed.

> I still think that Steve Jobs's important contributions could have been made without so many stories that we hear of his negative dealings with people.
>
> —Steve Wozniak[4]

At other times, he rationalizes Jobs's behavior as effective when instances that are perceived as negative do occur, even suggesting that Jobs possessed traits that General Colin Powell considered leadership qualities but perceived as negative when expressed by Jobs.

> He spends a long time thinking about products and directions in private, asking himself every question he can think of and formulating the answers. By the time he presents an idea, it is very well tested in his own mind and he has a big advantage over others. It's fair to call this a form of intelligence.
>
> —Steve Wozniak[5]

> I recently read a great set of points by General Colin Powell about leadership qualities of CEO's and it fit Steve perfectly. I guess a lot of CEOs see things in him that we call negatives as positives. It's partly an outcome of the capitalist system as to which personal traits do better in business.
>
> —Steve Wozniak[6]

CHRISANN BRENNAN

Jobs was subject to what he perceived as blackmail attempts, as he once feared in disclosures to the federal government. The attempts were not from a kidnapper, though—they were from the mother of Jobs's first child, Chrisann Brennan. In 2005, she decided that she should be compensated by Jobs for his actions related to their child born in 1978. Brennan estimated $25 million would make her feel better and provide closure on their relationship.

> What I want is money that will more than last me for the rest of my life. I believe that decency and closure can be achieved through money. It is very simple.
>
> —Chrisann Brennan[7]

In 2009, when their daughter was then 31, Brennan reportedly offered Jobs an ultimatum—provide for her financial well-being or she would publish her book about her relationship with Jobs, *A Bite in the Apple*.

> I don't react well to blackmail … I will have no part in any of this.
>
> —Steve Jobs, copying his adult daughter Lisa on the response[8]

ROLE MODELS AS INNOVATORS

Jobs did have role models in business, including Edwin Land, a founder of Polaroid who also left his own company. Land was much like Jobs in that he focused on connecting technology and the arts.

> The man is a national treasure. I don't understand why people like that can't be held up as models: This is the most incredible thing to be—not an astronaut, not a football player—but this.
>
> —Steve Jobs[9]

Land hired art history majors and had them take science courses, "in order to create chemists who could keep up when his conversation turned from Maxwell's equations to Renoir's brush strokes."[10]

> Every significant invention … must be startling, unexpected, and must come into a world that is not prepared for it. If

> the world were prepared for it, it would not be much of an invention.
>
> —Edwin Land[11]

Jobs shared Land's belief that innovations should be unexpected, even going so far as to suggest that consumers should never be asked what they wanted, because the consumer wouldn't be able to state what they wanted (and would want something else by the time the product was created). John Sculley affirmed that Jobs admired Dr. Land, saying both believed their products always existed, simply awaiting someone to discover them.

> Both of them had this ability to not invent products, but discover products. Both of them said these products have always existed—it's just that no one has ever seen them before. We were the ones who discovered them.
>
> —John Sculley[12]

Sculley also noted that Ross Perot visited Macintosh on a few occasions, and that Akio Morita—the man who built Sony—was another hero of Jobs.[13]

LACK OF KNOWN CHARITABLE INVOLVEMENT

While Jobs was perceived as not being involved in charitable support with his wealth accrued through multiple corporations, he did originally use personal—and Apple—resources to promote charity and provide computers for educational institutions.

PERSONAL INVOLVEMENT IN SEVA FOUNDATION

After meeting Dr. Larry Brilliant during his 1974 trip to India, Jobs would help Brilliant provide vision care to the poor in Nepal through Brilliant's Seva Foundation. As a page on Seva's website affirms:

> Steve participated as an advisor in Seva's first meetings, and is recognized as one of the organization's co-founders. He recognized the vital nature of data for program planning and quality

> assurance. His donation of one of the first Apple IIs and VisiCalc enabled Seva's U.S. and Nepali team to enter and analyze eye care survey results.[14]

In addition to his advice about outcomes and one of the first Apple II computers, Jobs provided a $5,000 check to Brilliant's start-up initiative, which is still providing services almost four decades later. And in helping to start Seva, he would meet individuals at the organizational meeting who had already influenced his life, whether during his brief time at Reed College or during the sojourn to India.

> On one side of the table was Ram Dass, the Jewish-born Hindu yogi who in 1971 had published one of Steve's favorite books, *Be Here Now* … Steve recognized a few of the folks. Robert Friedland, the guy who had convinced him to make a pilgrimage to India in 1974, came up and said hello.[15]

At that initial meeting, Jobs met the author of the book that had bound him to Dan Kottke, as well as the individual who had recommended he visit India. And while his was a comparatively small donation for someone of his wealth, it was a very impactful donation.

> I do want to counter the meme that he was disinterested in philanthropy and things for the greater good … It wasn't true.
>
> —Larry Brilliant[16]

Brilliant also stated that Jobs felt he could make a large impact on the world as a technology executive, as that's what he knew how to do best.

> I only know how to do one thing well … I think I can help the world by doing this one thing
>
> —Steve Jobs, per Larry Brilliant[17]

A START AND STOP TO PHILANTHROPY

Apple—through Jobs—was involved in extensive contributions to education before Jobs decided that computers alone might not solve the problems in education. Jobs separately claimed in 1985 that he had more money than he would ever spend in his life, and that just about everyone could spend money more wisely than government. He mentioned that he would start a foundation when he had time, but that giving away the money would be a challenge. And he mentioned that a measurement system and a method of seeking the most powerful ideas was important (and time-consuming).

> One is that in order to learn how to do something well, you have to fail sometimes. In order to fail, there has to be a measurement system. And that's the problem with most philanthropy—there's no measurement system. You give somebody some money to do something and most of the time you can really never measure whether you failed or succeeded in your judgment of that person or his ideas or their implementation. So if you can't succeed or fail, it's really hard to get better. Also, most of the time, the people who come to you with ideas don't provide the best ideas. You go seek the best ideas out, and that takes a lot of time.
>
> —Steve Jobs[18]

Jobs involvement in charity started on an encouraging note, then became completely private and/or nonexistent. Jobs's own Steven P. Jobs Foundation was founded in 1986 and closed in 1987, as per Andrew Ross Sorkin and Mark Vermilion, who was hired to run the foundation:

> In 1986, after leaving Apple and founding NeXT, he started the Steven P. Jobs Foundation. But he closed it a little over a year later. Mark Vermilion, whom Mr. Jobs hired away from Apple to run the foundation, said in an interview, "He clearly didn't have the time." Mr. Vermilion said that Mr. Jobs was interested in financing programs involving nutrition and vegetarianism, while Mr. Vermilion pushed him toward social entrepreneurism. "I don't know if it was my inability to get him excited about it," he said. "I can't criticize Steve."[19]

Not only did Jobs start and close his own foundation, he separately eliminated Apple's corporate charity programs upon returning in 1997, and didn't reinstate the program despite Apple accruing billions of dollars in stored cash. Sorkin further noted:

> But in 1997, when Mr. Jobs returned to Apple, he closed the company's philanthropic programs. At the time, he said he wanted to restore the company's profitability. Despite the company's $14 billion in profits last year [2010] and its $76 billion cash pile today, the giving programs have never been reinstated.[20]

Alterman felt Sorkin was far too soft on Jobs, as if Sorkin feared Jobs while he was still alive.

> A second issue raised by Jobs's life and death is all that money he accumulated. When *The New York Times* "DealBook" editor Andrew Ross Sorkin wrote a column before Jobs died, wondering why he seemed so stingy with his fortune—noting also that he did away with all the company's charity programs (which were restored after his departure in August)—Sorkin addressed the topic so gingerly, I half thought he feared Jobs would send a thunderbolt from the sky to disable his typing fingers.[21]

In fact, Apple resumed charitable efforts within a month of Jobs's resignation, with the announcement from new CEO Tim Cook coming less than three weeks after Jobs's resignation from Apple.

> Starting September 15 [2011], when you give money to a non-profit 501(c)(3) organization, Apple will match your gift dollar-for-dollar, up to $10,000 annually. This program will be for full-time employees in the US at first, and we'll expand it to other parts of the world over time.
>
> —Tim Cook[22]

Kahney astutely made a comparison noting that Bill Gates—seen as a rich monopolist with Microsoft—was intentional in improving the world through the Bill and Melinda Gates Foundation and The Giving Pledge, while Steve Jobs—the same age—was not publicly dedicated to making those form of visible improvements outside of running computer firms.

> In fact, the reality is reversed. It's Gates who's making a dent in the universe, and Jobs who's taking on the role of single-minded capitalist, seemingly oblivious to the broader needs of society.
>
> —Leander Kahney[23]

Bill Gates was changing the world in precisely the same manner Jobs claimed was required back in 1985; seeking measurable opportunities and spending in a few targeted areas for improvement that weren't seeing improvement through government initiatives; noting that a goal would be to give away all the money effectively rather than simply leaving the funds for his heirs.

> A lot of the people who are getting into philanthropy now are trying to put their smarts into it, their creativity into it, so they can change the way philanthropy is done. I don't get that feeling from him. I get the feeling that he's so into doing what he's doing that there's no creativity left. He's an artist, Steve. He either likes what he's looking at or he doesn't. He's not concerned with what contribution he's making. He wants to astound himself, for himself.
>
> —Philanthropist friend of Steve Jobs[24]

LIMITED EXCESSIVE CONSUMPTION

Jobs was known for very few ostentatious uses of his wealth, living well below the means for a billionaire, lacking even private security at his home. In 1995, he made a very strong declaration about his intended uses of his wealth in the future, stating:

> There's no yacht in my future … I've never done this for the money, but I'm grateful that people are believing in our dream and the reality of our achievements.
>
> —Steve Jobs[25]

And for much of his life, he lived with just a few, well-thought out possessions. But fifteen years later, there indeed was a yacht in Steve

Jobs's future, named *Venus*, at a cost of well over $130 million, with design ideas coming directly from Steve Jobs (which caused the yacht to be described as looking "strange for a boat").[26] Sadly, the yacht would be completed the year after his death, but friend Bill Gates noted:

> He showed me the boat he was working on and said he was looking forward to being on it, even though we both knew there was a good chance that wouldn't happen. Thinking about your potential mortality isn't very productive.
>
> —Bill Gates[27]

Jobs demonstrated a consistent desire for simplicity in his products and his life. He became known for his uniform of black mock turtlenecks, jeans, and sneakers, which were worn daily, even for Apple promotional events. Walter Isaacson finally coaxed the origination story of this uniform from Jobs; he had actually tried to create a uniform for Apple employees (but failed):

> Sony, with its appreciation for style, had gotten the famous designer Issey Miyake to create its uniform. It was a jacket made of rip-stop nylon with sleeves that could unzip to make it a vest. So Jobs called Issey Miyake and asked him to design a vest for Apple, Jobs recalled, "I came back with some samples and told everyone it would great if we would all wear these vests. Oh man, did I get booed off the stage. Everybody hated the idea."
>
> In the process, however, he became friends with Miyake and would visit him regularly. He also came to like the idea of having a uniform for himself, both because of its daily convenience (the rationale he claimed) and its ability to convey a signature style. "So I asked Issey to make me some of his black turtlenecks that I liked, and he made me like a hundred of them." Jobs noticed my surprise when he told this story, so he showed them stacked up in the closet. "That's what I wear," he said. "I have enough to last for the rest of my life."[28]

DESIGN AND ARCHITECTURE

John Sculley, the CEO with whom Jobs had a falling out that lasted from 1985 through the end of Jobs's life, described their initial bond in 1983 being over design elements of Sculley's home, although Jobs's home at the time was simplistic (effectively empty).

> Steve from the moment I met him always loved beautiful products, especially hardware. He came to my house and he was fascinated because I had special hinges and locks designed for doors. I had studied as an industrial designer and the thing that connected Steve and me was industrial design. It wasn't computing.
>
> —John Sculley[29]

> I remember going into Steve's house and he had almost no furniture in it. He just had a picture of Einstein, whom he admired greatly, and he had a Tiffany lamp and a chair and a bed. He just didn't believe in having lots of things around but he was incredibly careful in what he selected. The same thing was true with Apple.
>
> —John Sculley[30]

Jobs was passionate about the architecture of Apple and Pixar facilities, providing ideas on Apple facilities and designing the entire Pixar campus as he best saw fit.

> Steve knew that the best architecture comes from solving design problems in a very simple and straightforward way ... He was quite knowledgeable about architecture and design, and he would ask very pointed questions: Can we do this? Why not? And for the most part, his questions would take us into places we hadn't considered before.
>
> —Karl Backus, designer of most Apple Stores[31]

He was ostensibly less passionate about the architecture of homes he purchased. Jobs bought a historic mansion named Jackling House, lived there a decade, petitioned to tear down the house, left the home empty and open to the weather for years, offered to give the home to anyone who moved it, and then finally got approval to tear down the house in 2011, shortly before his death.

> The saga of Jobs and the 17,250-square-foot, 14-bedroom mansion dates back to 1984, when he purchased it for a reported

$2 million. He was 29 and had just launched the Macintosh. Early photos of Jobs at the house suggest its fine condition, though his bachelor furnishings apparently amounted to little more than a mattress and state-of-the-art audio system. Whatever his original intentions, he would later scorn the house as "an abomination." After living there a decade and then renting it out, he left it vacant from about 2000.

> I bought [the property] to tear down the house, [but] I've been very busy the last 20 years.
>
> —Steve Jobs[32]

NEW YORK ATHLETIC CLUB

Jobs possessed a membership in an athletic organization—the New York Athletic Club—that had discriminated against specific minorities, religions, and women, even after a U.S. Supreme Court ruling requiring integration of their membership; this was the only organization to which he belonged as of the early 1990s, as released in his 1991 FBI file. There was never any evidence that Jobs knew the organization was still excluding women despite the court rulings, or that he could even recognize that the club was restricted to male-only admissions policies, as he apparently never visited.

> He belonged to no organizations other than the New York Athletic Club, however he had never been in the New York Athletic Club and knew nothing with regard to their membership policies.[33]

The New York Athletic Club finally complied with the court order in July 1989, as "the last holdout of its kind in New York City, even after June 1988, when the United States Supreme Court unanimously upheld a 1984 New York City law aimed primarily at requiring the admission of women to large, private clubs that play an important role in business and professional life."[34]

Part 6

DEATH AND LEGACY

Before embarking, he'd looked at his sister Patty, then for a long time at his children, then at his life's partner, Laurene, and then over their shoulders past them.

Steve's final words were:

OH WOW. OH WOW. OH WOW.

—Steve Jobs's last words, as told by his sister Mona Simpson[1]

Chapter 14

DYING

When asked if he knew what he wanted to do with the rest of his life, the 29-year-old Jobs still employed by Apple responded:

> There's an old Hindu saying that comes into my mind occasionally: "For the first 30 years of your life, you make your habits. For the last 30 years of your life, your habits make you." As I'm going to be 30 in February, the thought has crossed my mind.
>
> —Steve Jobs, 1985[1]

REFERENCES TO DEATH

Throughout his life, Steve Jobs spoke of death in almost three decades of interviews and speeches. The death included his own, but also his customers, his products, and his companies. He realized that even a long lifetime was still comparatively short, and frequently spoke of doing what one could on each day of a lifetime.

Jobs only lived to the age of 56; at the age of 29, he was at the intersection of four years that constituted both the first and last 30 years of his life. And every few years afterwards, he would speak of death and or dying, even before his initial diagnosis of cancer.

> Well probably death is the best invention of life. Ah … because it means there's a constant turnover. And so if you want to make a change in our society, the best place to do it is in the educational system.
>
> —Steve Jobs, 1992[2]

> Life is short, and we're all going to die really soon. It's true, you know.
>
> —Steve Jobs, 1998[3]

> Remembering that I'll be dead soon is the most important tool I've ever encountered to help me make the big choices in life.
>
> —Steve Jobs, 2005[4]

> I mean, some people say, "Oh, God, if [Jobs] got run over by a bus, Apple would be in trouble."
>
> —Steve Jobs, 2008[5]

ON/OFF SWITCH

Jobs made a comment to his authorized biographer where he placed a 50/50 probability that there was/was not a God and afterlife. Then he joked that the concept of nothing after death might explain why he didn't like on/off switches.

> "He said, 'Sometimes I'm 50–50 on whether there's a God. It's the great mystery we never quite know. But I like to believe there's an afterlife. I like to believe the accumulated wisdom doesn't just disappear when you die, but somehow it endures.'"[6]
>
> Jobs paused for a second, remembers Isaacson.
>
> "And then he says, 'But maybe it's just like an on/off switch and click—and you're gone.' And then he paused for another second and he smiled and said, 'Maybe that's why I didn't like putting on/off switches on Apple devices.'"[7]

PREPARATIONS FOR DEATH

> *A doctor told Mr. Jobs that the pioneering treatments of the kind he was undergoing would soon make most types of cancer a manageable chronic disease. Later, Mr. Jobs told Mr. Isaacson that he was either going to be one of the first "to outrun a cancer like this" or be among the last "to die from it."*[8]

Steve Jobs began planning for his death more than a year in advance. Protecting the privacy of both himself and his wife, an initial move was placing his home and two other properties into two trusts in March 2009.

As the *Globe and Mail* reported after his death, there is little need for public disclosure if assets are placed into a trust. While a Last Will and Testament is ordinarily a record available to the public, the presence of a trust simplifies the process of recording and reporting.

> "In that case, Mr. Jobs's will would merely say that everything is left to the trustee," she said. The assets would be then administered according to the trust rules, which are almost always private.[9]

Estate planning was critically important to Jobs, as he had three children with Laurene Powell-Jobs plus a child with Chrisann Brennan. If anything were to be left for Lisa Brennan-Jobs or Chrisann Brennan, it would have to be specified in his will and/or trust.

> A child does not have an automatic right to share in the inheritance, so how much Jobs chose to leave her [Lisa]—and his other children—was up to him. When people don't make their intent clear in their estate planning documents, families often fight over their wishes, especially when there are children of different parents involved.[10]

Chrisann Brennan had asked Steve Jobs for money during his lifetime, and apparently did not receive any funds from his will or trust. Brennan would later elect to contact Jobs's widow Laurene (as she had done previously with Steve), asking for a financial settlement through his estate, without media involvement. Brennan would later release copies of the communications she had sent to the media.

> You are in a position to help me without harm to your own life situation and children... If you can find your way to helping so

> that I, as Lisa's mother, can live in dignity and peace, we don't need to tell anyone ... this could be very quietly and legally done.
>
> —Chrisann Brennan[11]

Brennan also managed to get herself removed from the guest list for his memorial service, after allowing an article of their pairing to be printed in *Rolling Stone* just one week after Jobs's death. Julia McKinnell noted in *Maclean's*: "She's been punished in the past whenever she's cooperated with the media. The most recent occasion was in 2011, after Jobs's death, when she granted permission to *Rolling Stone* to print a piece about her and Steve. The move got her uninvited to Jobs's memorial service at Stanford University."[12]

FINAL DAYS

In his final days, Jobs elected to meet with many individuals, including Walter Isaacson, John Lasseter (Disney/Pixar), Bill Gates (Microsoft), and Bill Clinton. Yet he did not reach out to one of his oldest friends.

> Three weeks before, John (Lasseter, of Pixar) had visited Steve for the last time. "We sat for about an hour talking about coming projects he was so interested in," John said, his voice catching. "I looked at him and I realized this man had given me—given us—everything that we could ever want. I gave him a big hug. I kissed him on the cheek and for all of you"—John was crying now—"I said, 'Thank you. I love you, Steve.'"
>
> —Edwin Catmull (of Pixar and Disney)[13]

Near the end, Bill Gates wrote a friendly letter to congratulate Jobs on work he had done, building multiple companies and a family: "After Jobs's death, Gates received a phone call from his wife, Laurene. She said; 'Look, this biography [Walter Isaacson's] really doesn't paint a picture of the mutual respect you had.' And she said he'd appreciated my letter and kept it by his bed."[14]

Former President Bill Clinton revealed a surprising statement of Jobs's battle with cancer.

> I went to see him a few months before he passed. The last time I talked to him, he said, "This cancer I have is clever. It keeps

coming up with new ways to attack me. I don't think I have any weapons left, but I had a good time trying to beat it."

—Bill Clinton[15]

Wozniak assumed that if Jobs were close to dying, he would receive a call. The call never came.

I actually sort of thought that if he had a very short time that was defined, he would call me … I never got a call like that.

—Steve Wozniak[16]

DEATH

Terminally ill, Steve Jobs's personality remained exacting. He selected the nurses that would care for him in his remaining time on earth. However, the selection process was a challenge, as many nurses did not meet his lofty standards. He went through more than 60 nurses before fully accepting the three who would care for him, who were acknowledged in his eulogy.

Even ill, his taste, his discrimination and his judgment held. He went through 67 nurses before finding kindred spirits and then he completely trusted the three who stayed with him to the end. Tracy. Arturo. Elham.

—Mona Simpson[17]

Steve Jobs died of respiratory arrest, with the metastatic pancreas neuroendocrine tumor as the fundamental underlying cause. As his death was expected, no autopsy was performed.[18] According to the official death certificate, he is buried at Alta Mesa Memorial Park in Palo Alto, which is also the resting place of David Packard; Jobs's first employer at age 12 was Hewlett-Packard.[19] There is no marking on Jobs's grave.

Chapter 15

LEGACY OF STEVE JOBS IN INTELLECTUAL PROPERTY

Steve Jobs's legacy in intellectual property is mixed; Apple had won the first lawsuit—all the way back in 1983—that confirmed software was subject to intellectual property rights. Apple varied between a hardware firm and a software firm over time, and Jobs was intimately involved in creating intellectual property, protecting intellectual property, fighting off claims of infringement (when he founded NeXT), and in terms of precluding competition through forms of anticompetitive behavior and collusion.

CREATION OF INTELLECTUAL PROPERTY

Steve Jobs is listed as an inventor/contributor on more than 350 issued patents in the United States.[1] These are mostly design patents—"the visual ornamental characteristics embodied in, or applied to, an article of manufacture"; how products look, rather than how they work.[2] Over 35 years in computing, this would come out to an average of 10 per year but there's a catch.

> Jobs was not listed on any patents issued during his initial tenure at Apple.

> Jobs was not listed on any patents issued during his time at NeXT.
> Jobs was not listed on any patents issued during his time at Pixar.

Every one of the patents listing him was issued after his return to Apple Computer, so the average is closer to 25 patents per year. While he likely did not derive each of the concepts that led to a patent, the extensive involvement in these design patents clearly contributes to the perception of Jobs as a master innovator.

PROTECTING INTELLECTUAL PROPERTY

Jobs realized that his firm's continued success required enforcing patents, copyrights, and trademarks assigned to Apple. That initial copyright win meant that firms spending millions of dollars in development could expect protections from the judicial system to prevent copying by competitors. Under Jobs, Apple continued to actively enforce their intellectual property, believing that many major competitors—including Google and Microsoft—were blatantly stealing their innovations.

In his 2007 announcement of the iPhone, Jobs was explicit in stating Apple had over 200 patents on the iPhone alone. Phone manufacturer HTC became Apple's proxy for Google, as HTC was using the Google Android operating system. After filing a lawsuit against HTC in 2010, Steve Jobs declared:

> We can sit by and watch competitors steal our patented inventions, or we can do something about it. We've decided to do something about it.
>
> —Steve Jobs[3]

The lawsuit against Samsung came next on April 15, 2011, as Apple deemed the Galaxy series of phones to infringe upon iPhone's patents.[4]

ANTICOMPETITIVE BEHAVIOR

While Bill Gates and Microsoft fought charges of anticompetitive behavior throughout the 1990s—including an order at one time to divide Microsoft into two different companies—Steve Jobs was accused of anti-competitive behavior in lawsuits that originated in the early 2000s and continued after his death. At times, his communications could even be perceived as veiled—or not so veiled—threats to competitors.

Chen noted a series of those e-mails that had been made public, like a 2003 e-mail that he sent to other Apple executives about Music-Match opening their own music store, which would be a competitor to the combination of iTunes and the iPod.

> We need to make sure that when Music Match launches their download music store they cannot use iPod ... Is this going to be an issue?
>
> —Steve Jobs[5]

Apple's iTunes product for Windows was released later in 2003, Music-Match was acquired by Yahoo! in 2004, and the service was closed in 2008.

Jobs also communicated with executives with other firms in an attempt to stop those firms from attempting to hire Apple's employees. He reached out to Eric Schmidt, who happened to be the CEO at Google in 2006. Coincidentally, Eric Schmidt was appointed to the Board of Directors at Apple in 2006.

> I am told that Googles [sic] new cellphone software group is relentlessly recruiting in our iPod group ... If this is indeed true, can you put a stop to it?
>
> —Steve Jobs[6]

In a repeat in March of 2007, Jobs reached out to Google's Schmidt again, which led to a rapid string of events within Google. As Ames compiled from court records, Jobs began:

> I would be very pleased if your recruiting department would stop doing this.—Steve Jobs[7]

Schmidt sent the following e-mail internally at Google the next morning, which was acted upon immediately.

> I believe we have a policy of no recruiting from Apple and this is a direct inbound request. Can you get this stopped and let me

> know why this is happening? I will need to send a response back to Apple quickly so please let me know as soon as you can.
>
> —Eric Schmidt, CEO of Google[8]

When told that the Google recruiter was fired within an hour, Steve Jobs sent the message with his commentary to Apple's head of HR. The content of his message was a simple smiley face: :)

In a later court deposition, Sergey Brin, one of the wealthiest individuals in the world as a co-founder of Google, inadvertently provided commentary about Jobs's personality and influence (and the desire of other technology magnates to be seen approvingly by Jobs).

> Wow, Steve used a smiley. God, I never got one of those.
>
> —Sergey Brin, co-founder of Google[9]

SANCTIONS FOR ANTICOMPETITIVE BEHAVIOR

Like Bill Gates and Microsoft, a series of firms—Adobe, Apple, Google, Intel, Intuit, and Pixar—were found to have violated the Sherman Antitrust Act. With the six firms, the violations were detailed in the High-Tech Employees Antitrust Litigation, where the companies' collusion restricted job opportunities for talented employees who happened to be employed by one of the other companies by formal agreement. Apple was involved in three of the five illegal agreements, stopping employees from going to/from positions with Adobe, Google, and Pixar.[10] While Apple was found to have illegally worked with at least Adobe, Google, and Pixar to stop the movement of employees among those firms, Steve Jobs also approached other technology executives and was denied. When an executive of Palm responded that he wouldn't play along, Steve Jobs threatened Palm with potential patent infringement lawsuits.

> Mr. Jobs also tried to make a no-poaching agreement with Palm. When a Palm executive rejected that idea in an email, Mr. Jobs replied, "My advice is to take a look at our patent portfolio before you make a final decision here."
>
> —Steve Jobs[11]

Jobs also saw the ability to profit from digital media, and Apple was accused of colluding with providers to uniformly raise prices—and profits—for every company selling electronic books. Amazon, the largest e-book provider at the time, had a policy of $9.99 pricing.

> Throw in with Apple and see if we can all make a go of this to create a real mainstream e-books market at $12.99 and $14.99.
>
> —Steve Jobs[12]

Jobs's attempts to set high prices for e-books led to a settlement that was over a half billion dollars. When announcing the products that would compete with providers like Amazon, he even declared that there would be no price difference between his company's new product and Amazon's. As *The New York Times* reported,

> The $400 million will be paid on top of earlier settlements with publishers in the case, which provided $166 million in damages for consumers of e-books.
>
> In one instance, Mr. Jobs made comments to a reporter after he introduced the iPad and the iBookstore in January 2010. When asked why consumers would buy an e-book from Apple's bookstore instead of Amazon.com, Mr. Jobs replied, "The prices will be the same."[13]

Apple—under Jobs—was making agreements with other companies to manipulate competition (in seeking employees and ensuring high prices for consumers) in the first decade of the 2000s.

Chapter 16

LEGACY IN THE INDUSTRY

After Steve Jobs died, Richard Stengel of *Time* magazine wrote "This is Steve's seventh TIME cover, which puts him in the category of presidents and other world leaders."[1] In fact, Stengel underestimated the iconic nature of Steve Jobs; he had appeared on the cover of *Time* magazine eight times in the span of 29 years (and more impressively, seven times in the preceding fourteen years).[2]

ACCOLADES

Immediately after his passing, Jobs received accolades across the spectrum of technology firms, animation companies, and politicians. And he uniformly received praise for his accomplishments, whether from his own firm or competitors who were spurred on and improved based upon the innovations he was able to spearhead to market.

APPLE

Steve's greatest contribution and gift is the company and its culture. He cared deeply about that. He put in an enormous amount of time designing the concept for our new campus: That was a gift to the next generation. Apple University is another example of that. He wanted to use it to grow the next generation of leaders at Apple, and to make sure the lessons of the past weren't forgotten.

—Tim Cook[3]

My closest and my most loyal friend.

—Jony Ive[4]

Steve Jobs worked on a very different personal operating system than most people… It would be comparable to, say, getting a fish to fly.

—Guy Kawasaki, former Apple Marketing Director[5]

Steve was really good at bringing out the best in people, in getting them to reach beyond what they even knew they could do.

—Unnamed Apple Manager[6]

DREAMWORKS

The Thomas Edison of our time. In the way that Edison affected so many businesses, so did Steve Jobs.

—Jeffrey Katzenberg, Chairman and CEO of DreamWorks[7]

MICROSOFT

The integrated approach works when Steve is at the helm. But it doesn't mean it will win many rounds in the future.

—Bill Gates[8]

ORACLE

He [Jobs] was our Edison, he was our Picasso. He was an incredible inventor.

—Larry Ellison[9]

Larry Ellison believed Apple would not continue on the same upward trajectory without Jobs.[10]

> He's irreplaceable. Yeah. They—I don't see how they—how they can—how they can— they will not nearly so successful because he's gone … we conducted the experiment. I mean, it's been done. We saw Apple with Steve Jobs. We saw Apple without

> Steve Jobs. We saw Apple with Steve Jobs. Now, we're gonna see Apple without Steve Jobs.
>
> —Larry Ellison[11]

FORMER APPLE CEOS

In 2010, Sculley had described his feeling that Jobs was the only person who could save Apple; he elevated that praise after Jobs's death. Despite his own time leading Pepsi and Apple, Sculley himself would have the opportunity to declare Jobs to be "the greatest CEO ever." Not the greatest CEO of Apple, but the greatest CEO of all time.

> Steve Jobs taught us many, many lessons, and he was brilliant, but the reality is most of us aren't Steve Jobs ... You've got to assemble an incredibly great team, and what most people overlook with him and I know, because I was with him, is that he was brilliant at being able to recruit talent. And he did it by his charismatic ability to tell a compelling story with metaphors and poetry in ways that got people to do things they never thought they were capable of.
>
> —John Sculley[12]

LEGACY AT APPLE

Steve Jobs announced his resignation from Apple on August 24, 2011 and died October 5, 2011. In late September 2014, Tim Cook provided information about the former office of the iconic co-founder, when discussing the new headquarters then under construction.

> His office is still left as it was on the fourth floor, his name is still on the door.
>
> —Tim Cook[13]

As the CEO successor to Jobs, Cook had to implement the lessons from "the "best teacher I ever had by far"[14] and lead what he referred to as "Steve's company"; even four years later, he was describing the company as such:

> This is Steve's company. This is still Steve's company. It was born that way, it's still that way. And so his spirit I think will always be the DNA of this company.
>
> —Tim Cook[15]

While Cook still felt the company belonged to Steve, he noted:

> I loved Steve. Steve is not my competition ... he selected me. I want to do every single thing I can do and use every ounce of energy I've got to do as well as I can.
>
> —Tim Cook[16]

But just like Jobs was still present ...

> It's a bar of excellence that merely good isn't good enough. It has to be great. As Steve used to say, "Insanely great."
>
> —Tim Cook[17]

The legacy of Steve Jobs's influence on the culture and success of Apple can now be debated—the pipeline of new project development under Jobs was approximately two- and-a-half years.

IN THE MEDIA

The mass-market, best-selling authorized biography of Steve Jobs also garnered criticism. Isaacson's last interview with Jobs was a few weeks before his death, yet the book was expedited into print after Jobs's death on October 5, 2011, printed and shipped by October 24, 2011.

> So it's no surprise that Walter Isaacson's new biography of Apple founder and serial business inventor Steve Jobs, rushed into print less than a month after Jobs's death on October 5, is ultimately disappointing. Jobs is as much a mystery on the last page as he is on the first. Even those who loved or hated him the most can't quite make up their minds about him; Isaacson makes sure to let us know that Jobs's friends and family consistently acknowledged his flaws, while his opponents (Bill Gates leaps to mind) felt compelled to praise his consistent pattern of game-changing business invention.[18]

Multiple Apple executives felt the depiction of Jobs in Isaacson's book was inaccurate, with at least one stating that the portrayal in a much later book, *Becoming Steve Jobs*, was far more appropriate.

> But now its account is being challenged by another book, this one called *Becoming Steve Jobs*, by Brent Schlender, a veteran

> technology journalist who was friendly with Jobs, and Rick Tetzeli, executive editor at *Fast Company*. Some of Jobs's former colleagues and friends have taken sides, speaking out against the old book and praising the new one. Tim Cook, Apple's CEO and Jobs's successor, has said that Isaacson's book depicts Jobs as "a greedy, selfish egomaniac." Jony Ive, Apple's design chief, has weighed in against it, and Eddy Cue, Apple's vice president of software and Internet services, tweeted about the new book: "Well done and first to get it right."
>
> Jobs was a man of towering contradictions: he identified deeply with the counterculture but spent his life in corporate boardrooms amassing billions; he made beautiful products that ostensibly enabled individual creativity but in their architecture expressed a deep-seated need for central control.[19]

In visual media, depictions of Jobs were often designed to capture viewers, rather than tell his story. Steinberg lamented that the 2015 *Steve Jobs* movie was told through three product launches. Ignored entirely were Pixar, his family, and the illness over a period of eight years that ended up claiming his life.

> The film spends little time on topics that don't serve its theme. There's nothing about Pixar, the animation company Mr. Jobs built, or his marriage to Laurene Powell and their family. There is no mention of his battle with cancer.[20]

LEGACY

Steve Jobs is also recalled in more contemporary discussions about decision making. Mark Hurd was fired by Hewlett-Packard (and then hired by Larry Ellison at Oracle). Ellison not only hired Hurd, but made Hurd co-CEO of Oracle after stepping down from day-to-day responsibilities, referring to his previous firing at Hewlett-Packard as the

> Worst personnel decision since the idiots on the Apple board fired Steve Jobs many years ago.
>
> —Larry Ellison[21]

Jobs has influenced future innovators and entrepreneurs. Even Chinese entrepreneurs mirror Steve Jobs and cite his actions. Jia Yueting of Leshi Internet Information & Technology (LeTV) releases products wearing blue jeans, gray sneakers, and black t-shirts, and even ran a modified "1984" ad where his firm took on *Apple*. And when Jia's Board of Directors disagree with him, he does what he feels Jobs would do (and Jobs did with NeXT and Pixar)—he starts a new company with his own funds.

> That's why Apple was successful ... When Jobs came back, he had power over his board. His charisma convinced everybody he was right.
>
> —Jia Yueting[22]

WHAT CAN WE LEARN FROM JOBS IN BUSINESS?

The story of Steve Jobs is a story of intuition. He always took a product with the idea of making it accessible to the greatest number of people. From assembling kit computers to marketing one of the first computers that could simply be plugged in, to releasing a computer that had a visual interface rather than a series of text commands. The descriptions in his words—and others—was that he wanted to make technology products so simple that they would be as intuitive to use as a toaster or telephone.

> Business schools often ask me what Steve Jobs teaches us about leadership ... It's that he took total responsibility for his products from end to end, that he put products above return on investment and that he wasn't a slave to focus groups.
>
> —Walter Isaacson[23]

In getting other people to iteratively derive the simplest, most elegant, artistic experience for those customers, he was demanding, frustrating and hurtful to many individuals. Not everyone could tolerate sharing the journey of making the "insanely great" products with Jobs.

> Jobs built a good team, but he would have gotten even better people if he'd been less tough on them.
>
> —Mike Useem, Wharton Leadership Center[24]

CONCLUSION

In a period of less than 57 years, Steve Jobs showed that an individual from humble beginnings, grateful just for being born, could impact society and a major industry. He saw opportunities where others did not, and over time he would force what he perceived to be those opportunities on others within his businesses.

Raised by Paul and Clara Jobs, he had all he ever wanted in childhood. Yet in his early days as a parent, he was an absentee father like his own biological father. And worse, he denied his parentage for a substantial period of time. His relationship with his first child was not always smooth, but Jobs married and developed a devoted family life with his next three children. That was just one way he evolved over time.

Intellectually gifted, he wasn't the best student. Yet he was a voracious reader and routinely incorporated cultural and artistic references into his work. He started college, only completed one semester for credit, yet was one of the most forward-thinking individuals in society. His vision for his firms was the vision for his firms.

As a teenager, he was involved in illegal activities such as hacking phone systems and using illegal drugs; while he learned from those experiences, he also was fortunate that he never suffered legal ramifications for those actions. He would later credit his illegal acts with inspiring some of the products his company would create later.

During his teenage years and early in adulthood, he understood how to make money quickly. He was a hustler, using the skills of a good friend for business purposes the friend had not contemplated. A primary reason we have easy-to-use technological devices—computers, phones, and tablets—was his ability to see what a customer of average technical sophistication would need in order to intuitively use a device.

He learned that even in a company he helped co-found, he could be removed without careful placement of individuals in the company. From there, he created another company and helped a lagging company reach a much higher potential. The company he founded was NeXT, which we see manifested in operating systems even today. The company he rescued was a division of Lucasfilm that became the wildly successful Pixar, the leader in computer-based animation.

He had one more company to rescue. Coming back to an Apple hemorrhaging cash, he returned the firm to profitability quickly. Even before coming back to Apple, he had admitted to a television host that he knew how to fix an Apple that had been lacking innovation in the decade-plus period he was gone. Focusing on the consumer market other manufacturers had abandoned during a period of exceptionally low computer prices, Apple became—at one point—the most valuable company in the history of the world. And from that period, electronics consumers received intuitive user interfaces, services, and devices that will be sold for many years hence, including iTunes, iPod, iPhone, and iPad.

Jobs's legacy lies in the ideas he brought to the consumer market and the products made possible by those ideas, whether made by Apple and Pixar or a competitor firm. His legacy lies in the culture of Apple and Pixar, which had been configured to continue in his absence. And his legacy lies in his family, shaped by their interactions with Jobs and with the financial resources to determine how each may want to make a future mark on the world.

Appendix
HONORS AND AWARDS

1985 National Medal of Technology
Steven P. Jobs and Steve Wozniak
Apple Computer, Inc.
For their development and introduction of the personal computer which has sparked the birth of a new industry extending the power of the computer to individual users.[1]

1987 Samuel S. Beard Award for Greatest Public Service 35 Years or Under.[2]

1989 *Inc.* magazine's "Entrepreneur of the Decade"' *Inc.* magazine notes that "it was not until we had settled on him as our Entrepreneur of the Decade that we realized how little we actually knew about him."[3]

2007 California Hall of Fame[4]
The criteria for selection to the California Hall of Fame requires that inductees have transcended the boundaries of their field to make lasting contributions to the state, nation and world and that their extraordinary vision motivates and inspires people to further their own dreams.[5]

2012 Grammy Trustees Award

Special Merit Award is presented by vote of The Recording Academy's National Trustees to individuals who, during their careers in music, have made significant contributions, other than performance, to the field of recording.[6]

2013 Disney Legend

He was an early investor and chief executive of Pixar, and became the Walt Disney Company's largest shareholder overnight when it acquired Pixar Animation Studios in 2006. That same year, he joined the Disney board of directors, and remained a valuable sounding board and advisor to the company until his passing in 2011.[7]

NOTES

CHAPTER 1

1. Steve Lohr (January 12, 1997). "Creating Jobs," *The New York Times*. http://www.nytimes.com/1997/01/12/magazine/creating-jobs.html

2. Daniel Morrow (April 20, 1995). "Oral History Interview with Steve Jobs," *Smithsonian Institution*. http://americanhistory.si.edu/collections/comphist/sj1.html

3. David Sheff (February 1985). "Playboy Interview: Steve Jobs," *Playboy*.

4. CBS News (October 23, 2011). Steve Jobs: Revelations from a Tech Giant.

5. Jon Brooks (November 25, 2011). Interview: Apple Employee No. 12 Dan Kottke on Company's Earliest Days and the College Steve Jobs. KQED. http://ww2.kqed.org/news/programs/news-fix.

6. Tom Junod (October 2008). "Steve Jobs and the Portal to the Invisible," *Esquire*.

7. Steve Lohr (January 12, 1997). "Creating Jobs," *The New York Times*. http://www.nytimes.com/1997/01/12/magazine/creating-jobs.html

8. Ibid.

9. Tom Junod (October 2008). "Steve Jobs and the Portal to the Invisible," *Esquire*.

10. Daniel Morrow (April 20, 1995). "Oral History Interview with Steve Jobs," *Smithsonian Institution*. Ibid.

11. Ibid.

12. Ibid.

13. Alan Deutschman (2000). *The Second Coming of Steve Jobs*. Random House.

14. Jason Hiner (December 2014) "Apple's First Employee: The Remarkable Odyssey of Bill Fernandez," *Tech Republic*. http://www.techrepublic.com/article/apples-first-employee-the-remarkable-odyssey-of-bill-fernandez/

15. Ibid.

16. Steve Wozniak. http://archive.woz.org/letters/general/71.html Set 71

17. David Sheff (February 1985). "Playboy Interview: Steve Jobs," *Playboy*.

18. Walter Isaacson (2011). *Steve Jobs*. Simon & Schuster.

19. Esquire Editors (October 15, 2015). "How Blue Box Phone Phreaking Put Steve Jobs and Woz on the Road to Apple," *Esquire*.

20. Robert X. Cringely (1996). *PBS*. Triumph of the Nerds.

21. Steve Wozniak (2000). http://archive.woz.org/letters/general/59.html

22. Steve Wozniak and Gina Smith (2006). W.W. Norton. iWoz: Computer Geek to Cult Icon.

23. Owen W. Linzmayer (2004). *Apple Confidential 2.0: The Definitive History of the World's Most Colorful Company*. No Starch Press.

24. Dag Spicer (2002). Computer History Museum. Steve Jobs: From Garage to World's Most Valuable Company. http://www.computerhistory.org/atchm/steve-jobs/#disqus_thread

25. Robert X. Cringely (1996). *PBS*. Triumph of the Nerds.

26. Geeta Dayal (February 1, 2013) The Slate Book Review. Phreaks and Geeks: Before Steve Jobs and Steve Wozniak invented Apple, They Hacked Phones. http://www.slate.com/articles/technology/books/2013/02/steve_jobs_and_phone_hacking_exploding_the_phone_by_phil_lapsley_reviewed.2.html

27. Department of Defense (n.d.). Steve Jobs's DoD Background Check.

28. John Markoff (October 5, 2011). "Steve Jobs of Apple Dies at 56," *The New York Times*.

29. Mike Cassidy (October 10, 2011). "Cassidy: Woz and Jobs Were Silicon Valley's Biggest Buddy Story." *San Jose Mercury News*. http://www.mercurynews.com/ci_19082667

30. Federal Bureau of Investigation (n.d.). Steve Jobs's FBI File.

CHAPTER 2

1. Owen W. Linzmayer (2004). *Apple Confidential 2.0: The Definitive History of the World's Most Colorful Company*. No Starch Press.

2. Jay Cocks (January 3, 1983). "The Updated Book of Jobs," *Time*.

3. Andrew B. Wilson and Robert O. Skovgard (December 2011 / January 2012). "Steve Jobs vs. Bill Gates," *American Spectator*.

4. Jon Brooks (November 25, 2011). Interview: Apple Employee No. 12 Dan Kottke on Company's Earliest Days and the College Steve Jobs. KQED. http://ww2.kqed.org/news/programs/news-fix.

5. Chris Lydgate (December 2011). "In Memoriam: Steven P. Jobs '76, Prodigal Son," *Reed Magazine*. http://www.reed.edu/reed_magazine/in-memoriam/obituaries/december2011/steve-jobs-1976.html

6. Charles Cubeta (March 2, 2012). "Bowdoin's 'Toxic' Son," *The Bowdoin Orient*. http://bowdoinorient.com/article/7094

7. Jon Brooks (November 25, 2011). *KQED*. Interview: Apple Employee No. 12 Dan Kottke on Company's Earliest Days and the College Steve Jobs. KQED. http://ww2.kqed.org/news/programs/news-fix.

8. Andrew B. Wilson and Robert O. Skovgard (December 2011 / January 2012). "Steve Jobs vs. Bill Gates,"*American Spectator*.

9. Jon Brooks (November 25, 2011). Interview: Apple Employee No. 12 Dan Kottke on Company's Earliest Days and the College Steve Jobs. KQED. http://ww2.kqed.org/news/programs/news-fix.

10. Henry Adams (October 5, 2011). "A Tribute to a Great Artist: Steve Jobs," *Smithsonian.com*. http://www.smithsonianmag.com/arts-culture/a-tribute-to-a-great-artist-steve-jobs-99783256/?no-ist

11. Chris Lydgate (December 2011). "In Memoriam: Steven P. Jobs '76, Prodigal Son," *Reed Magazine*. http://www.reed.edu/reed_magazine/in-memoriam/obituaries/december2011/steve-jobs-1976.html

12. Steve Jobs (August 27, 1991). "Staying Hungry," as printed in Sallyportal. http://www.reed.edu/reed_magazine/sallyportal/posts/2013/staying-hungry.html

13. TNN (October 7, 2011). Trip to India as teen was a life-changer for Steve Jobs. http://articles.economicTimes.indiaTimes.com/2011-10-07/news/30253986_1_steve-jobs-story-of-apple-computer-steve-wozniak

14. Nick Wingfield (November 20, 2013). "A Gift from Steve Jobs Returns Home," *The New York Times*. http://bits.blogs.ny*Times*.com/2013/11/20/a-gift-from-steve-jobs-returns-home/

15. Annie Gowen (October 31, 2015). "Inside the Indian Temple that Draws America's Tech Titans," *The Washington Post*. https://www.washingtonpost.com/world/asia_pacific/inside-the-indian-temple-that-draws-americas-tech-titans/2015/10/30/03b646d8-7cb9-11e5-bfb6-65300a5ff562_story.html

16. Brent Schlender and Rick Tetzeli (2015). *Becoming Steve Jobs*. Crown Business.

17. Walter Isaacson (October 29, 2011). "The Genius of Jobs," *The New York Times*. http://www.nyTimes.com/2011/10/30/opinion/sunday/steve-jobss-genius.html

18. Annie Gowen (October 31, 2015). Inside the Indian Temple that Draws America's Tech Titans," *The Washington Post*. https://www.washingtonpost.com/world/asia_pacific/inside-the-indian-temple-that-draws-americas-tech-titans/2015/10/30/03b646d8-7cb9-11e5-bfb6-65300a5ff562_story.html

19. John Markoff (2005). *What the Dormouse Said: How the Sixties Counter Culture Shaped the Personal Computer Industry*. Penguin Books.

20. Department of Defense (n.d.). Steve Jobs's DoD Background Check.

21. Federal Bureau of Investigation (n.d.). Steve Jobs's FBI File.

22. Julia McKinnell (November 4, 2013). "Steve Jobs: Genius, and Lousy Father," *Maclean's*.

23. Owen W. Linzmayer (2004). *Apple Confidential 2.0: The Definitive History of the World's Most Colorful Company*. No Starch Press.

24. Mike Cassidy (March 28, 2013). "Cassidy on Nolan Bushnell: 'Steve was difficult,' Says Man Who First Hired Steve Jobs," *San Jose Mercury News*. http://www.mercurynews.com/ci_22890892/cassidy-steve-jobs-hire-nolan-bushnell-book-atari

25. Steve Wozniak (2000). http://archive.woz.org/letters/general/91.html

26. Owen W. Linzmayer (2004). *Apple Confidential 2.0: The Definitive History of the World's Most Colorful Company*. No Starch Press.

27. Steve Wozniak (2000). http://archive.woz.org/letters/general/22.html

28. Andrew Marszal (December 14, 2011). "Steve Wozniak: Steve Jobs' Deal Made Me Cry," *The Telegraph*. http://www.telegraph.co.uk/technology/news/8956282/Steve-Wozniak-Steve-Jobs-deal-made-me-cry.html

29. Steve Wozniak (2000). http://archive.woz.org/letters/general/91.html

30. Steve Wozniak (n.d.). http://www.woz.org/letters?page=19

31. Steve Wozniak (2000). http://archive.woz.org/letters/general/19.html

32. Department of Defense (n.d.). Steve Jobs's DoD Background Check.

33. Barclay Walsh (February 25, 2016). What Do Jerry Brown, Steve Jobs and Kamala Harris Have in Common? *All Things Considered (NPR)*.

34. Owen W. Linzmayer (2004). *Apple Confidential 2.0: The Definitive History of the World's Most Colorful Company*. No Starch Press.

35. Michael Becraft (2014). *Bill Gates: A Biography*. ABC-CLIO.

36. Steve Wozniak (2000). http://archive.woz.org/letters/general/41.html

37. Owen W. Linzmayer (2004). *Apple Confidential 2.0: The Definitive History of the World's Most Colorful Company*. No Starch Press.

38. Ibid.

CHAPTER 3

1. Robert X. Cringely (1996). *PBS*. Triumph of the Nerds.

2. Andrew Marszal (December 14, 2011). "Steve Wozniak: Steve Jobs' Deal Made Me Cry," *The Telegraph*. http://www.telegraph.co.uk/technology/news/8956282/Steve-Wozniak-Steve-Jobs-deal-made-me-cry.html

3. Malcolm Gladwell (2008). *Outliers: The Story of Success*. Hachette Book Group.

4. Jason Hiner (December 2014). "Apple's First Employee: The Remarkable Odyssey of Bill Fernandez," *Tech Republic*. http://www.techrepublic.com/article/apples-first-employee-the-remarkable-odyssey-of-bill-fernandez/

5. David B. Yoffie and Michael A. Cusumano (2015). *Strategy Rules: Five Timeless Lessons from Bill Gates, Andy Grove, and Steve Jobs*. Harper Collins.

6. Owen W. Linzmayer (2004). *Apple Confidential 2.0: The Definitive History of the World's Most Colorful Company*. No Starch Press.

7. Jason Green (October 29, 2013). "Steve Jobs' Childhood Home Becomes a Landmark," *San Jose Mercury News*. http://www.mercurynews.com/news/ci_24410143/steve-jobs-childhood-home-becomes-a

8. Jason Green (September 25, 2013). "Steve Jobs' Sister Weighs In on Effort to Preserve Apple Co-founder's Childhood Home in Los Altos," *San Jose Mercury News*. http://www.mercurynews.com/breaking-news/ci_24169962/jobs-sister-weighs-effort-preserve-apple-co-founders

9. Jeff Goodell (April 4, 1996). "The Rise and Fall of Apple Inc.," *Rolling Stone*.

10. Norman Sklarewitz (April 1, 1979). "From the Very Beginning, Apple Was Born to Grow," *Inc Magazine*. http://www.Inc.com/magazine/1979/04/apple-born-to-grow_pagen_3.html

11. Steve Wozniak (2000). Letters-General Questions Answered. https://web.archive.org/web/20000817234627/http://www.woz.org/letters/pirates/09.html

12. Owen W. Linzmayer (2004). *Apple Confidential 2.0: The Definitive History of the World's Most Colorful Company*. No Starch Press.

13. Johnny Dodd (December 10, 2014). "Apple Co-Founder Auctions Off Archives to Supplement His Social Security Income," *People*. http://www.people.com/article/ron-wayne-apple-co-founder-auctions-archives

14. Ibid.

15. Owen W. Linzmayer (2004). *Apple Confidential 2.0: The Definitive History of the World's Most Colorful Company*. No Starch Press.

16. *WGBH* (May 14, 1990). Interview with Steve Jobs. http://openvault.wgbh.org/catalog/7b7ae3-steve-jobs-interview

17. Ken Gagne (August 26, 2011). "Steve Wozniak on Steve Jobs' Resignation," *Computerworld*. http://www.computerworld.com/article/2470933/windows-pcs/steve-wozniak-on-steve-jobs--resignation.html

18. Ibid.

19. Jon Zilber (Spring 2015). "Silicon Valley's Merry Prankster Put His Degree on Hold and Reshaped the World," *California Magazine*. http://alumni.berkeley.edu/california-magazine/spring-2015-dropouts-and-drop-ins/silicon-valley-s-merry-prankster-put-his

20. Jon Brooks (December 8, 2011). Interview: Apple Employee No. 12 Dan Kottke on the Macintosh, Jobs and Woz, Stock Options (Pt 2). *KQED*. http://ww2.kqed.org/news/2011/12/08/interview-apple-employee-no-12-dan-kottke-on-the-early-days-jobs-and-woz-stock-options-pt-ii

21. Steve Wozniak (2000). http://archive.woz.org/letters/general/17.html

22. Steve Wozniak (2000). http://archive.woz.org/letters/general/66.html

23. George Gendron and Bo Burlingham (April 1, 1989). "The Entrepreneur of the Decade," *Inc.* http://www.Inc.com/magazine/19890401/5602.html

24. *WGBH* (May 14, 1990). Interview with Steve Jobs. http://openvault.wgbh.org/catalog/7b7ae3-steve-jobs-interview

25. Steve Wozniak (2000). http://archive.woz.org/letters/general/13.html

26. Jason Green (September 25, 2013). "Steve Jobs' Sister Weighs In on Effort to Preserve Apple Co-founder's Childhood Home in Los Altos," *San Jose Mercury News*. http://www.mercurynews.com/breaking-news/ci_24169962/jobs-sister-weighs-effort-preserve-apple-co-founders

27. Steve Wozniak (2000). http://archive.woz.org/letters/general/68.html Set 68

CHAPTER 4

1. *WGBH* (May 14, 1990). Interview with Steve Jobs. http://openvault.wgbh.org/catalog/7b7ae3-steve-jobs-interview

2. Stephen G. Wozniak (April 11, 1977). US Patent 4136359. Microcomputer for use with video display.

3. Steve Wozniak (2000). http://archive.woz.org/letters/general/109.html

4. Steve Wozniak (2000). http://archive.woz.org/letters/general/8.html

5. Steve Wozniak (2000). http://archive.woz.org/letters/general/36.html

6. Jon Zilber (Spring 2015). "Silicon Valley's Merry Prankster Put His Degree on Hold and Reshaped the World," *California Magazine*. http://alumni.berkeley.edu/california-magazine/spring-2015-dropouts-and-drop-ins/silicon-valley-s-merry-prankster-put-his

7. Steve Wozniak (2000). http://archive.woz.org/letters/general/23.html

8. Norman Sklarewitz (April 1, 1979). "From the Very Beginning, Apple Was Born to Grow," *Inc Magazine*. http://www.inc.com/magazine/1979/04/apple-born-to-grow_pagen_3.html

9. Jim Edwards (December 26, 2013). "These Pictures of Apple's First Employees Are Absolutely Wonderful, *Business Insider*.

10. Norman Sklarewitz (April 1, 1979). "From the Very Beginning, Apple Was Born to Grow," *Inc Magazine*. http://www.inc.com/magazine/1979/04/apple-born-to-grow_pagen_3.html

11. Robert X. Cringely (1996). *PBS*. Triumph of the Nerds.

12. Harvard Business School (2001). Interview with Arthur Rock. http://www.hbs.edu/entrepreneurs/pdf/arthurrock.pdf

13. Ibid.

14. Jay Yarow (May 24, 2011). "EXCLUSIVE: Interview with Apple's First CEO Michael Scott," *Business Insider*.

15. Ibid.

16. Ibid.

17. Michael Krantz (October 18, 1999) "Steve Jobs at 44," *Time*. http://www.time.com/Time/printout/0,8816,32207,00.html

18. Jay Yarow (May 24, 2011). "EXCLUSIVE: Interview with Apple's First CEO Michael Scott," *Business Insider*.

19. Ibid.

20. Norman Sklarewitz (April 1, 1979). "From the Very Beginning, Apple Was Born to Grow," *Inc Magazine*. http://www.inc.com/magazine/1979/04/apple-born-to-grow_pagen_3.html

21. Robert X. Cringely (1996). "Accidental Empires," *Harper Business*.

22. Michael Becraft (2014). *Bill Gates: A Biography*. ABC-CLIO.

23. *Macworld* (November 2011). "Steve Jobs: The Man Who Saved Apple."

24. Steve Wozniak (2000). http://archive.woz.org/letters/general/56.html

25. Daniel Morrow (April 20, 1995). Smithsonian Institution. Oral History Interview with Steve Jobs. http://americanhistory.si.edu/collections/comphist/sj1.html

26. *WGBH* (May 14, 1990). Interview with Steve Jobs. http://openvault.wgbh.org/catalog/7b7ae3-steve-jobs-interview

27. Robert X. Cringely (1996). *PBS*. Triumph of the Nerds.

28. Jeff Goodell (June 16, 1994). "Steve Jobs in 1994," *Rolling Stone*. http://www.rollingstone.com/culture/news/steve-jobs-in-1994-the-rolling-stone-interview-20110117?print=true

29. Andrew B. Wilson and Robert O. Skovgard (December 2011 / January 2012). "Steve Jobs vs. Bill Gates," *American Spectator*.

30. Michael I. Hyman (1995). *PC Roadkill*. John Wiley & Sons Inc.

31. Steve Wozniak (2000). http://archive.woz.org/letters/general/68.html

32. Kara Swisher and Walt Mossberg (May 31, 2007). "Bill Gates and Steve Jobs at D5," *AllThingsD*.

33. Ibid.

34. Chrisann Brennan (2013). *The Bite in the Apple*. St. Martin's Press.

35. Ibid.

36. Ibid.

37. Ibid.

38. George Gendron and Bo Burlingham (April 1, 1989). "The Entrepreneur of the Decade," *Inc*. http://www.Inc.com/magazine/19890401/5602.html

CHAPTER 5

1. Robert X. Cringely (1996). *PBS*. Triumph of the Nerds.

2. Owen W. Linzmayer (2004). *Apple Confidential 2.0: The Definitive History of the World's Most Colorful Company*. No Starch Press.

3. Apple (2016). Apple—Investor Relations. http://investor.apple.com/faq.cfm

4. Vijith Assar (August 16, 2013). "Early Apple Employees Talk Memories of Steve Jobs, New Movie," *Vice*. http://insights.dice.com/2013/08/16/early-apple-employees-talk-memories-of-steve-jobs-new-movie-2/

5. National Transportation Safety Board (NTSB). (n.d.) NTSB Identification: LAX81FA044. http://www.ntsb.gov/_layouts/ntsb.aviation/brief.aspx?ev_id=27749&key=0

6. Steve Wozniak (2000). http://archive.woz.org/letters/general/51.html

7. Valerie Rice (April 15, 1985). "Unrecognized Apple II Employees Exit," InfoWorld.

8. Michael Becraft (2014). *Bill Gates: A Biography*. ABC-CLIO.

9. Walter Isaacson (2011). *Steve Jobs*. Simon & Schuster.

10. Ibid.

11. Jay Cocks (January 3, 1983). "The Updated Book of Jobs," *Time*.

12. Ibid.

13. Ibid.

14. Ibid.

15. Michael Moritz (2009). *Return to the Little Kingdom: Steve Jobs, the Creation of Apple, and How It Changed the World*. Overlook Press.

16. Bill Gates and Steve Ballmer (1983). Microsoft Applications Strategy Memo.

17. David Allison (1993). Transcript of a Video History Interview with Mr. William "Bill" Gates. http://americanhistory.si.edu/comphist/gates.htm

18. *Apple Computer, Inc. v. Franklin Computer Corporation* (August 30, 1983). U.S. Court of Appeals Third Circuit. http://digital-lawonline.info/cases/219PQ113.htm

CHAPTER 6

1. Steve Jobs (2005). Stanford Commencement Address.

2. Robert X. Cringely (1996). *PBS*. Triumph of the Nerds.

3. Ibid.

4. Barbara Chai (October 6, 2015). "Speakeasy: John Sculley on Steve Jobs: The Movie, the Man and His Own Role in the Drama," *The Wall Street Journal*. http://blogs.wsj.com/speakeasy/2015/10/06/john-sculley-on-steve-jobs-the-movie-the-man-and-his-own-role-in-the-drama/

5. Jon Brooks (December 8, 2011). Interview: Apple Employee No. 12 Dan Kottke on the Macintosh, Jobs and Woz, Stock Options

(Pt 2). *KQED*. http://ww2.kqed.org/news/2011/12/08/interview-apple-employee-no-12-dan-kottke-on-the-early-days-jobs-and-woz-stock-options-pt-ii

6. Walter Isaacson (2011). *Steve Jobs*. Simon & Schuster.

7. Contactmusic (January 6, 2012). "Joan Baez Pays Tribute to Generous Steve Jobs." http://www.contactmusic.com/joan-baez/news/joan-baez-pays-tribute-to-generous-steve-jobs_1281311

8. Jerry Manock (n.d.). "Invasion of Texaco Towers," *Folklore.org*. http://www.folklore.org/StoryView.py?project=Macintosh&story=Invasion_of_Texaco_Towers.txt

9. Joan Baez (1987). *And a Voice to Sing With: A Memoir*. Summit Books.

10. Chris Blackhurst (December 18, 2015). "From Fergie to Steve Jobs, a Billionaire Venture Capitalist's Leadership Lessons," *Evening Standard*.

11. Andy Hertzfeld (n.d.) "Signing Party." Folklore.org: http://www.folklore.org/StoryView.py?story=Signing_Party.txt

12. Steve Jobs (1983). Apple Keynote.

13. Steve Hayden (January 30, 2011). "'1984': As Good as It Gets," *Advertising Age*.

14. Ibid.

15. Robert X. Cringely (1996). *PBS*. Triumph of the Nerds.

16. Harry McCracken (January 25, 2014). "Exclusive: Watch Steve Jobs' First Demonstration of the Mac for the Public, Unseen Since 1984," *Time*. http://time.com/1847/steve-jobs-mac/

17. Tom Reilly (November 21, 2011). "How Steve Jobs and the Invention of the Mac Saved My Life," *Out*.

18. Tom Zito (Fall 1984). "The Bang Behind the Bucks, the Life Behind the Style," *Newsweek*.

19. Ibid.

20. Ibid.

21. David Sheff (February 1985). "Playboy Interview: Steve Jobs," *Playboy*.

22. Ibid.

23. Ibid.

24. Ibid.

25. Owen W. Linzmayer (2004). *Apple Confidential 2.0: The Definitive History of the World's Most Colorful Company*. No Starch Press.

26. John Markoff (October 5, 2011). "Steve Jobs of Apple Dies at 56," *The New York Times*.

27. Chris Preimesberger (November 8, 2013). "eWeek at 30: Steve Jobs Returns in 1997 to Revive a Moribund Apple," *eWeek*.

28. Barbara Rudolph, Robert Buderi, and Karen Horton (September 30, 1985). "Shaken to the Very Core: After Months of Anger and Anguish, Steve Jobs Resigns as Apple Chairman," *Time*.

29. Ibid.

30. Ibid.

31. Ibid.

32. Jim Edwards (May 27, 2015). "Former Apple CEO John Sculley Admits Steve Jobs Never Forgave Him, and He Never Repaired their Friendship, before Jobs Died," *Business Insider*. http://www.businessinsider.com/john-sculley-admits-steve-jobs-never-forgave-him-before-jobs-died-2015-5?r=UK&IR=T

33. Gerald Lubenow and Michael Rogers (September 29, 1985). "Jobs Talks about His Rise and Fall," *Newsweek*.

34. Ibid.

35. Ibid.

36. Ibid.

CHAPTER 7

1. George Gendron and Bo Burlingham (April 1, 1989). "The Entrepreneur of the Decade," *Inc*. http://www.Inc.com/magazine/19890401/5602.html

2. Philip Elmer-DeWitt, Jonathan Beaty, and Thomas McCarroll (June 20, 1988). "The Case of the Missing Machine What in the World is Steve Jobs Doing, and When Will He Do It?" *Time*.

3. *The New York Times* (January 30, 1987). "Company News; Perot Reportedly Invests in Next."

4. Philip Elmer-DeWitt, Jonathan Beaty, and Thomas McCarroll (June 20, 1988). "The Case of the Missing Machine What in the World is Steve Jobs Doing, and When Will He Do It?" *Time*.

5. George Gendron and Bo Burlingham (April 1, 1989). "The Entrepreneur of the Decade," *Inc*. http://www.Inc.com/magazine/19890401/5602.html

6. *WGBH* (May 14, 1990). Interview with Steve Jobs. http://openvault.wgbh.org/catalog/7b7ae3-steve-jobs-interview

7. Brian Dumaine and Rosalind Klein Berlin (October 18, 1993). "America's Toughest Bosses," *Fortune*. http://archive.fortune.com/magazines/fortune/fortune_archive/1993/10/18/78470/index.htm

8. Ibid.

9. Ibid.

10. Ilan Mochari (June 9, 2015). "Steve Jobs Revered This Designer's Dictatorial Approach," *Inc.* http://www.inc.com/ilan-mochari/design-paul-rand.html?cid=readmore

11. George Gendron and Bo Burlingham (April 1, 1989). "The Entrepreneur of the Decade," *Inc.* http://www.Inc.com/magazine/19890401/5602.html

12. Ned Zeman (October 8, 1990). "A Case of Vapor?" *Newsweek*.

13. Newsweek Staff (November 17, 1991). "At Last: Signs of Life at NeXT," *Newsweek*.

14. Jeff Goodell (June 16, 1994). "Steve Jobs in 1994," *Rolling Stone*. http://www.rollingstone.com/culture/news/steve-jobs-in-1994-the-rolling-stone-interview-20110117?print=true

15. Ibid.

16. Ibid.

17. Ibid.

18. Ibid.

19. Daniel Morrow (April 20, 1995). "Oral History Interview with Steve Jobs," *Smithsonian Institution*.

20. Matt Deatherage (December 2011). "What Steve Jobs Did," *Macworld*.

21. Adam Rogers (September 4, 1995). "In Search of a Sequel," *Newsweek*.

22. Gary Wolf (February 1996). "Steve Jobs: The Next Insanely Great Thing," *Wired*. http://www.wired.com/wired/archive//4.02/jobs.html?topic=&topic_set=

CHAPTER 8

1. Cynthia Jensen (June 15, 2015). "Becoming Steve Jobs: The Evolution of a Reckless Upstart into a Visionary Leader," *Library Journal*.

2. Alvy Ray Smith (n.d.). Pixar Founding Documents. http://alvyray.com/pixar/default.htm

3. Julie Bort (June 4, 2014). "Steve Jobs Taught This Man How to Win Arguments with Really Stubborn People," *Business Insider*.

4. Ed Catmull and Amy Wallace (2014). *Creativity, Inc.: Overcoming the Unseen Forces That Stand in the Way of True Inspiration*. Random House.

5. Ibid.

6. Ibid.

7. Steve Lohr (January 12, 1997). "Creating Jobs," *The New York Times*. http://www.nytimes.com/1997/01/12/magazine/creating-jobs.html

8. *Newsweek* (December 11, 1995). "Comeback Kid."

9. Michael G. Gray (July 14, 1996). "On the Right Path?" *SF Gate*. http://www.sfgate.com/business/article/On-the-right-path-3135567.php

10. John Lasseter (December 26, 2011). "Steve Jobs," *Time*.

11. Richard Verrier and Claudia Eller (February 2, 2004). "A Clash of CEO Egos Gets Blame in Disney-Pixar Split," *Los Angeles Times*.

12. Ibid.

13. Laura M. Holson (January 25, 2006). "Disney Agrees to Acquire Pixar in a $7.4 Billion Deal," *The New York Times*. http://www.nytimes.com/2006/01/25/business/25disney.html?_r=1&oref=slogin&mtrref=undefined&gwh=6313CA318BC0688519DA20016AD27775&gwt=pay

PART 4

1. Michael Meyer (July 28, 1997). "A Death Spiral?" *Newsweek*.

2. *Macworld* (November 201). "Steve Jobs: The Man Who Saved Apple."

3. Brent Schlender (November 9, 1998). "The Three Faces of Steve," *Fortune*. http://money.cnn.com/magazines/fortune/fortune_archive/1998/11/09/250880/index.htm

CHAPTER 9

1. Steve Lohr (January 12, 1997). "Creating Jobs," *The New York Times*. http://www.nytimes.com/1997/01/12/magazine/creating-jobs.html

2. Ibid.

3. *Newsweek* (December 30, 1996). "The Man Who was Next."

4. Chris Preimesberger (November 8, 2013). "eWeek at 30: Steve Jobs Returns in 1997 to Revive a Moribund Apple," *eWeek.*

5. John Simons (July 29, 1996). "A Bushel of Hope for Apple." *U.S. News & World Report.*

6. Ibid.

7. Seybold (February 1997). "Amelio Shoots Straight with Apple Shareholders," *Seybold Report on Internet Publishing.*

8. John Simons (February 17, 1997). "The Difference a Year Didn't Make," *U.S. News & World Report.*

9. Jeff Walsh (April 21, 1997). "Apple Anticipates Turning Profit by Year's End," *InfoWorld.*

10. Paul Sloan (November 17, 2009). "Perspective: Yahoo's Turnaround Efforts Are Straight from Steve Jobs' Playbook.," *CBS News MoneyWatch.* http://www.cbsnews.com/news/perspective-yahoos-turnaround-efforts-are-straight-from-steve-jobs-playbook/

11. Jim Louderback (April 6, 1998). "Amelio's Book Depicts a Bruised Apple," *PC Week.*

12. Robert X. Cringely (August 25, 1997). "If You're Still in the Dark about Politics or Apple, Then You're Not Alone," *InfoWorld.*

13. Brent Schlender (November 9, 1998). "The Three Faces of Steve," *Fortune.* http://money.cnn.com/magazines/fortune/fortune_archive/1998/11/09/250880/index.htm

14. Stephanie Gruner (September 1997). "The Most-Admired Entrepreneurs," *Inc.*

15. *Maclean's* (July 21, 1997). "Upsetting the Applecart."

16. Michael Meyer (July 28, 1997). "A Death Spiral?" *Newsweek.*

17. *InfoWorld* (August 4, 1997). "Rumors about Apple and Jobs Continue."

18. *Newsweek* (August 11, 1997). "Will He or Won't He?"

19. Steven Levy and Katie Hafner (August 18, 1997). "A Big Brother?" *Newsweek.*

20. Susan Gregory Thomas and Leonard Wiener (August 18, 1997). "Why Bill Gates and Steve Jobs Made Up," *U.S. News & World Report.*

21. *San Jose Mercury News* (November 11, 1997). "Steve Jobs Answers Questions on Apple's Future."

22. Michael Meyer (August 18, 1997). "Just Who'd Want This Job?" *Newsweek.*

23. Maryann Alioto (Fall 1997). "Lawrence Ellison," *Directors & Boards.*

24. Henry Norr (January 6, 2000). "MacWorld Expo/Permanent Jobs/Apple CEO Finally Drops 'Interim' from Title," *San Francisco Chronicle.*

25. Kara Swisher and Walt Mossberg (May 31, 2007). "Bill Gates and Steve Jobs at D5," *AllThingsD.*

26. Cathy Booth (August 18, 1997). "Steve's Job: Restart Apple," *Time Magazine.*

27. Steven Levy and Katie Hafner (August 18, 1997). "A Big Brother?" *Newsweek.*

28. Cathy Booth (August 18, 1997). "Steve's Job: Restart Apple," *Time Magazine.*

29. Rick Webb (November 12, 2011). *Betabeat.* Caught in the Webb: Let's Not Party Like It's 1999.

30. Cathy Booth (August 18, 1997). "Steve's Job: Restart Apple," *Time Magazine.*

31. Microsoft (August 6, 1997). Microsoft and Apple Affirm Commitment to Build Next Generation Software for Macintosh.

32. Cathy Booth (August 18, 1997). "Steve's Job: Restart Apple," *Time Magazine.*

33. Michael Meyer (July 28, 1997). "A Death Spiral?" *Newsweek.*

34. Susan Gregory Thomas and Leonard Wiener (August 18, 1997). "Why Bill Gates and Steve Jobs Made Up." *U.S. News & World Report.*

35. Steven Levy and Katie Hafner (August 18, 1997). "A Big Brother?" *Newsweek.*

36. *San Jose Mercury News* (November 11, 1997). "Steve Jobs Answers Questions on Apple's Future."

37. Ibid.

38. Bradley Johnson (July 21, 1997). "Jobs Dives into Apple Agency Review Process," *Advertising Age.*

39. Bradley Johnson (October 6, 1997). "Apple's Jobs Rehires Olivo to Head Up Advertising," *Advertising Age.*

40. Rita Clifton (2009). "Brands and Branding, 2nd Edition," *The Economist.*

41. Ibid.

42. Benj Edwards (December 2012). "Steve Jobs's Seven Key Decisions," *Macworld.*

43. Steven Levy and Katie Hafner (August 18, 1997). "A Big Brother?" *Newsweek.*

44. Ian Parker (February 23, 2015). "The Shape of Things to Come," *New Yorker.*

45. Chloe Albanesius (June 2012). "Apple Designer Jonathan 'Jony' Ive Awarded Knighthood," *PC Magazine.*

46. Ian Parker (February 23, 2015). "The Shape of Things to Come," *New Yorker.*

47. Ibid.

48. Seybold (October 13, 1997). "Jobs, Gates Highlight San Francisco Show," *Seybold Report on Publishing Systems.*

49. Leander Kahney (September 5, 2011). "The 10 Commandments of Steve," *Newsweek.*

50. Jillian D'Onfro (March 22, 2015). "Why Execs from Other Companies Wanted to Meet with Steve Jobs on Fridays," *Business Insider.* http://www.businessinsider.com/steve-jobs-at-pixar-versus-apple-2015-3

51. Seybold (April 1998). "Jobs Stirs Faithful Once Again," *Seybold Report on Internet Publishing.*

52. Andy Reinhardt (May 12, 1998). "Steve Jobs on Apple's Resurgence," *Business Week.* http://www.businessweek.com/bwdaily/dnflash/may1998/nf80512d.htm

53. Alyson Shontell (September 12, 2014). "Business Insider. How Steve Jobs Convinced Tim Cook to Work for Apple," *Business Insider.* http://www.businessinsider.com/how-steve-jobs-convinced-tim-cook-to-work-for-apple-2014-9

54. Brent Schlender (November 9, 1998). "The Three Faces of Steve," *Fortune.* http://money.cnn.com/magazines/fortune/fortune_archive/1998/11/09/250880/index.htm

55. Ibid.

56. Ibid.

57. Michael Krantz (October 18, 1999). "Steve Jobs at 44," *Time.* http://www.time.com/time /printout/0,8816,32207,00.html

58. Ibid.

59. Stephen F. Nathans (April 1999). "The Man Who Sold the World," *EMedia Professional.*

60. Steven Berglas (October 1999). "What You Can Learn from Steve Jobs," *Inc.*

61. Ibid.

62. Fortune Editors (October 7, 2011). "Noah Wyle on Playing Steve Jobs," *Fortune.* http://fortune.com/2011/10/07/noah-wyle-on-playing-steve-jobs/

63. Linda Bridges (March 1, 1999). "The 15 Most Influential," *PC Week.*

64. Michael Krantz (October 18, 1999). "Steve Jobs at 44," *Time.* http://www.time.com/time /printout/0,8816,32207,00.html

65. Michael Krantz and Janice Maloney (August 2, 1999). "Jobs' Golden Apple," *Time.*

66. Ibid.

67. *Fortune* (January 24, 2000). "Apple's One-Dollar-A-Year Man." http://money.cnn.com/magazines/fortune/fortune_archive/2000/01/24/272277/index.htm

68. Ibid.

69. *San Jose Mercury News* (February 21, 2000). "Analysts Disagree on Accounting for Apple Computer CEO's Jet."

70. Steve Wozniak (2000). http://archive.woz.org/letters/general/44.html

CHAPTER 10

1. *Macworld* (November 2011). "Steve Jobs: The Man Who Saved Apple."

2. Ibid.

3. Michael Becraft (2014). *Bill Gates: A Biography.* ABC-CLIO.

4. *Macworld* (November 2011). "Steve Jobs: The Man Who Saved Apple."

5. Josh Quittner (February 5, 2003). "Apple's New Core," *Time.*

6. Laura Locke (April 29, 2003). "Steve Jobs on the iTunes Music Store," *Technologizer.* http://technologizer.com/2011/12/07/steve-jobs-on-the-itunes-music-store-the-unpublished-interview/

7. Jeff Goodell (December 25, 2003). "Steve Jobs: *Rolling Stone*'s 2003 Interview," *Rolling Stone*. http://www.rollingstone.com/music/news/steve-jobs-rolling-stones-2003-interview-20111006

8. Ibid.

9. Apple Inc. (2005). iTunes Music Store Sells One Million Videos in Less Than 20 Days. http://www.apple.com/pr/library/2005/10/31iTunes-Music-Store-Sells-One-Million-Videos-in-Less-Than-20-Days.html

10. Steve Lohr (October 20, 2011). "Jobs Tried Exotic Treatments to Combat Cancer, Book Says," *The New York Times*. http://www.nytimes.com/2011/10/21/technology/book-offers-new-details-of-jobs-cancer-fight.html

11. Benny Evangelista (August 2, 2004). "Apple's Jobs has Cancerous Tumor Removed: He'll Take a Month Off to Recuperate," *SF Gate*. http://www.sfgate.com/news/article/Apple-s-Jobs-has-cancerous-tumor-removed-He-ll-2736823.php

12. Kara Swisher and Walt Mossberg (May 31, 2007). "Bill Gates and Steve Jobs at D5," *AllThingsD*.

13. Ibid.

14. Ibid.

15. Ibid.

16. Steven Levy (October 16, 2006). "Good for the Soul," *Newsweek*. http://www.msnbc.msn.com/id/15262121/site/*Newsweek*/print/1/displaymode/1098/

17. Apple Inc. (2006). Apple Inc. (AAPL) 10-K filed 12/29/2006.

18. Apple Inc. (2006). Apple's Special Committee Reports Findings of Stock Option Investigation. https://www.apple.com/pr/library/2006/10/04Apples-Special-Committee-Reports-Findings-of-Stock-Option-Investigation.html

19. Matt MacInnis (September 2014). "What I Learned from Steve Jobs," *Inc*.

20. John Lasseter (December 26, 2011). "Steve Jobs," *Time*.

21. Michael Becraft (2014). *Bill Gates: A Biography*. ABC-CLIO.

22. Walter Isaacson (2011). *Steve Jobs*. Simon & Schuster.

23. Michael Becraft (2014). *Bill Gates: A Biography*. ABC-CLIO.

24. Ryan Block (January 9, 2007). "Live from Macworld 2007: Steve Jobs Keynote," *Engadget*. http://www.engadget.com/2007/01/09/live-from-macworld-2007-steve-jobs-keynote/

25. Victor Luckerson (September 14, 2015). "Steve Jobs Totally Dissed the Stylus 8 Years Before Apple Pencil," *Time.com*.

26. Ryan Block (January 9, 2007). "Live from Macworld 2007: Steve Jobs Keynote," *Engadget*. http://www.engadget.com/2007/01/09/live-from-macworld-2007-steve-jobs-keynote/

27. Ibid.

28. David Lieberman (April 30, 2007). *USA Today*. CEO Forum: Microsoft's Ballmer Having a "Great" Time.

29. Fred Vogelstein (October 4, 2013). "And Then Steve Said, 'Let There Be an iPhone,'" *The New York Times Magazine*. http://www.nytimes.com/2013/10/06/magazine/and-then-steve-said-let-there-be-an-iphone.html

30. Ibid.

31. Ibid.

32. Ibid.

33. *CNBC* (September 5, 2007). Interview Transcript: Steve Jobs. http://www.cnbc.com/id/20613051

34. Peter Elkind (March 5, 2008). "The Trouble with Steve Jobs," *Fortune*. http://fortune.com/2008/03/05/the-trouble-with-steve-jobs/

35. Betsy Morris (February 2008). "Steve Jobs Speaks Out," *Fortune*. http://money.cnn.com/galleries/2008/fortune/0803/gallery.jobsqna.fortune/index.html

36. Ibid.

37. Ibid.

38. *Automotive News* (November 9, 2015). "Apple Car Was Already Planted in Steve Jobs's Mind."

39. Joe Nocera (July 26, 2008). "Apple's Culture of Secrecy," *The New York Times*. http://www.nytimes.com/2008/07/26/business/26nocera.html?pagewanted=all

40. Ibid.

41. *CBS News* (October 23, 2011). Steve Jobs: Revelations from a Tech Giant.

42. Ibid.

43. Steve Jobs (January 5, 2009). Letter from Apple CEO Steve Jobs.

44. Yukari Iwatani Kane and Joann S. Lublin (June 20, 2009). "Jobs Had Liver Transplant," *The Wall Street Journal.* http://www.wsj.com/articles/SB124546193182433491

45. Chris Gayomali (March 12, 2015). "Apple CEO Tim Cook Tried to Give Steve Jobs His Liver—But Jobs Refused," *Fast Company.* http://www.fastcompany.com/3043628/apple-ceo-tim-cook-tried-to-give-steve-jobs-his-liver-but-jobs-refused

46. Marc Perrusquia (December 5, 2013). "Southern Transplants: How Apple CEO Steve Jobs Got the Liver He Needed in Memphis," *The Memphis Commercial Appeal.* http://www.commercialappeal.com/news/southern-transplants-how-apple-ceo-steve-jobs-got-the-liver-he-needed-in-memphis-ep-307208860-329010091.html

47. Leander Kahney (October 14, 2010). "John Sculley on Steve Jobs," *Cult of Mac.* http://www.cultofmac.com/63295/john-sculley-on-steve-jobs-the-full-interview-transcript/

48. Ibid.

49. Ibid.

50. Ibid.

51. Ibid.

52. Doug Gross (June 25, 2010). *CNN.* "Apple on iPhone Complaints: You're Holding It Wrong." http://www.cnn.com/2010/tech/mobile/06/25/iphone.problems.response/

53. Mark Milian (November 23, 2011). *CNN.* "Steve Jobs Fielded Some Customer Support Requests." http://www.cnn.com/2011/11/22/tech/innovation/jobs-excerpt-customer-service/index.html

54. Joshua Gans (December 13, 2010). "The Steve Jobs Theory of Customer Relations," *Harvard Business Review.*

55. http://www.cbsnews.com/news/bill-gates-joins-the-ipads-armyof-critics-steve-jobs-couldnt-care-less/

56. James Koch, Robert Fenili, and Richard Cebula (March 2011). "Do Investors Care If Steve Jobs Is Healthy?" *Atlantic Economic Journal.*

57. Steve Jobs (August 24, 2011). Letter from Steve Jobs. http://www.apple.com/pr/library/2011/08/24Letter-from-Steve-Jobs.html

58. Brent Schlender and Rick Tetzeli (2015). *Becoming Steve Jobs.* Crown Business.

CHAPTER 11

1. Walter Isaacson (2011). *Steve Jobs*. Simon & Schuster.

2. Joan Frawley Desmond (Fall 2011). "Steve Jobs: An Unwanted Child," *Human Life Review*.

3. Ameena Meer (Summer 1987). "Mona Simpson," *Bomb*. http://bombmagazine.org/article/947/mona-simpson

4. Steve Lohr (January 12, 1997). "Creating Jobs," *The New York Times*.

5. John C. Abell (October 10, 2011). Steve Jobs' Biological Father: Some Contact (Maybe), But No Facetime," *Wired*.

6. Alexandra Berzon (October 10, 2011). "For Jobs' Biological Father, the Reunion Never Came," *The Wall Street Journal*. http://www.wsj.com/articles/SB10001424052970203499704576620911395191694

7. Post Staff Report (August 27, 2011). "Dad Waits for Jobs to iPhone," *New York Post*.

8. Ibid.

9. Walter Isaacson (2011). *Steve Jobs*. Simon & Schuster.

10. Mona Simpson (1986). *Anywhere but Here*. Knopf Doubleday Publishing Group.

11. Steve Lohr (January 12, 1997). "Creating Jobs," *The New York Times*. http://www.nytimes.com/1997/01/12/magazine/creating-jobs.html

12. Ameena Meer (Summer 1987). "Mona Simpson," *Bomb*. http://bombmagazine.org/article/947/mona-simpson

13. Tom Junod (October 2008). "Steve Jobs and the Portal to the Invisible," *Esquire*.

14. Mona Simpson (October 30, 2011). "A Sister's Eulogy for Steve Jobs," *The New York Times*. http://www.nytimes.com/2011/10/30/opinion/mona-simpsons-eulogy-for-steve-jobs.html?_r=0

15. Richard Appel (2005). 20th Century Fox. The Simpsons Complete Seventh Season DVD.

16. Mona Simpson (October 30, 2011). "A Sister's Eulogy for Steve Jobs," *The New York Times*. http://www.nytimes.com/2011/10/30/opinion/mona-simpsons-eulogy-for-steve-jobs.html?_r=0

17. Steve Wozniak (2000). http://archive.woz.org/letters/general/72.html

18. Steve Silberman (October 26, 2015). *PLOS*. "What Kind of Buddhist Was Steve Jobs, Really?"

19. Ibid.

20. Brent Schlender and Rick Tetzeli (2015). *Becoming Steve Jobs*. Crown Business.

21. Mark Milian (October 7, 2011). *CNN*. The Spiritual Side of Steve Jobs. http://edition.cnn.com/2011/10/05/tech/innovation/steve-jobs-philosophy/?hpt=ibu_c1

22. Tom Junod (October 2008). "Steve Jobs and the Portal to the Invisible," *Esquire*.

23. Tristan Quinn (December 14, 2011). "Steve Jobs: Billion Dollar Hippy BBC Two," *The Telegraph*. http://www.telegraph.co.uk/culture/tvandradio/8953371/Steve-Jobs-Billion-Dollar-Hippy-BBC-Two.html

24. Mona Simpson (October 30, 2011). "A Sister's Eulogy for Steve Jobs," *The New York Times*. http://www.nytimes.com/2011/10/30/opinion/mona-simpsons-eulogy-for-steve-jobs.html?_r=0

25. Jessica E. Lessin and Miriam Jordan (May 16, 2013). "Laurene Powell Jobs Goes Public to Promote Dream Act," *The Wall Street Journal*. http://www.wsj.com/articles/SB10001424127887323582904578487263583009532

26. Colleen Curry (October 6, 2011). *ABC News*. Steve Jobs Kept Private Life Closely Guarded Secret. http://abcnews.go.com/Technology/steve-jobs-secret-private-life/story?id=14678496

27. Ned Zeman (March 24, 1991). "A Jobs Wedding," *Newsweek*.

28. Connie Guglielmo (May 7, 2012). "A Day in the Life of Steve Jobs," *Forbes*. http://www.Forbes.com/sites/connieguglielmo/2012/05/07/a-day-in-the-life-of-steve-jobs/

29. Jessica E. Lessin and Miriam Jordan (May 16, 2013). "Laurene Powell Jobs Goes Public to Promote Dream Act," *The Wall Street Journal*. http://www.wsj.com/articles/SB10001424127887323582904578487263583009532

30. Lisa Brennan-Jobs (Spring 1999). "Driving Jane," *The Harvard Advocate*.

31. Lisa Brennan-Jobs (February 2008). "Tuscan Holiday," *Vogue*.

32. Ibid.

33. Ibid.

34. Gary Wolf (February 1996). "Steve Jobs: The Next Insanely Great Thing," *Wired*. http://www.wired.com/wired/archive//4.02/jobs.html?topic=&topic_set=

35. Nick Bilton (September 10, 2014). "Steve Jobs Was a Low-Tech Parent," *The New York Times*.

36. Rick Tetzeli (March 18, 2015). "Tim Cook on Apple's Future: Everything Can Change Except Values," *Fast Company*. http://www.fastcompany.com/3042435/steves-legacy-tim-looks-ahead

37. Peter Lattman and Claire Cain Miller (May 17, 2013). "Steve Jobs's Widow Steps onto Philanthropic Stage," *The New York Times*.

38. Margaret Kadifa (October 29, 2015). "Halloween at Steve Jobs's House," *The Houston Chronicle*.

39. Steven Levy and Katie Hafner (August 18, 1997). "A Big Brother?" *Newsweek*.

40. *CBS Interactive* (July 15, 2012). Steve Jobs: Family photo Album http://www.cbsnews.com/news/steve-jobs-family-photo-album/

CHAPTER 12

1. Tom Zito (Fall 1984). "The Bang Behind the Bucks, The Life Behind the Style," *Newsweek*.

2. Daniel Morrow (April 20, 1995). Oral History Interview with Steve Jobs, Smithsonian Institution. http://americanhistory.si.edu/collections/comphist/sj1.html

3. Ibid.

4. Gary Wolf (February 1996). "Steve Jobs: The Next Insanely Great Thing," *Wired*. http://www.wired.com/wired/archive//4.02/jobs.html?topic=&topic_set=

5. *Alberta Report* (February 12, 1996). "Vouchers, not Computers."

6. Paul Wells (October 24, 2011). "Freedom to Fail is What Made Steve Jobs," *Maclean's*.

7. Department of Defense (n.d.). Steve Jobs's DoD Background Check.

8. Ibid.

9. U.S. Department of Commerce, International Trade Administration (n.d). Charter of the President's Export Council. http://www.trade.gov/pec/charter.asp

10. Federal Bureau of Investigation (n.d.) Steve Jobs's FBI File.

11. Ibid.

12. Ibid.

13. Ibid.

14. Ibid.

15. Ibid.

16. Ibid.

17. Tom Brokaw and Jim Miklaszewski (February 25, 1997). *NBC Nightly News*. President Clinton Embroiled in Fundraising Scandal. https://static.nbclearn.com/files/nbcarchives/site/pdf/2967.pdf

18. Allie Townsend (October 11, 2011). "Chicago Ideas Week: What's in Bill Clinton's Head," *Time*. http://content.time.com/Time/specials/packages/article/0,28804,2096504_2096506_2096669,00.html

19. Philip Elmer-Dewitt (December 18, 2013). "Video: President Bill Clinton Remembers Apple's Steve Jobs," *Fortune Magazine*. http://fortune.com/2013/12/18/video-president-bill-clinton-remembers-apples-steve-jobs/

20. Walter Isaacson (2011). *Steve Jobs*. Simon & Schuster.

21. Eric Alterman (November 28, 2011). "Steve Jobs: An American 'Disgrace,'" *Nation*.

CHAPTER 13

1. Steve Wozniak (2000). http://archive.woz.org/letters/general/13.html

2. Steve Wozniak (2000). http://archive.woz.org/letters/general/27.html

3. Steve Wozniak (2000). http://archive.woz.org/letters/general/74.html

4. Steve Wozniak (2000). http://archive.woz.org/letters/general/85.html

5. Steve Wozniak (2000). http://archive.woz.org/letters/general/45.html

6. Steve Wozniak (2000). http://archive.woz.org/letters/general/61.html

7. Chrisann Brennan (December 2005). Letter to Steve Jobs. https://fortunedotcom.files.wordpress.com/2015/08/chrisann-letter.pdf

8. Peter Elkind (April 6, 2015). "When Steve Jobs's Ex-girlfriend Asked Him to Pay $25 Million for his 'Dishonorable behavior,'" *Fortune*. http://Fortune.com/2015/08/06/steve-jobs-apple-girlfriend/

9. Christopher Bonanos (October 7, 2011). "The Man Who Inspired Jobs," *The New York Times*. http://www.nytimes.com/2011/10/07/opinion/the-man-who-inspired-jobs.html?_r=0

10. Ibid.

11. Ibid.

12. Leander Kahney (October 14, 2010). "John Sculley on Steve Jobs," *Cult of Mac*. http://www.cultofmac.com/63295/john-sculley-on-steve-jobs-the-full-interview-transcript/

13. Ibid.

14. Seva Foundation. (n.d.) Steve Jobs & Seva. http://www.seva.org/site/PageServer?pagename=remembering_Steve_Jobs#.VvM0D4-cHIU

15. Brent Schlender and Rick Tetzeli (2015). *Becoming Steve Jobs*. Crown Business.

16. Nick Wingfield (November 20, 2013). "A Gift from Steve Jobs Returns Home," *The New York Times*. http://bits.blogs.nytimes.com/2013/11/20/a-gift-from-steve-jobs-returns-home/

17. Ibid.

18. David Sheff (February 1985). "Playboy Interview: Steven Jobs," *Playboy*. http://www.playboy.com/magazine/playboy-interview-steve-jobs

19. Andrew Ross Sorkin (August 29, 2011). "Dealbook: The Mystery of Steve Jobs's Public Giving," *The New York Times*. http://dealbook.nyTimes.com/2011/08/29/the-mystery-of-steve-jobss-public-giving/?_r=0

20. Ibid.

21. Eric Alterman (November 28, 2011). "Steve Jobs: An American 'Disgrace,'" *Nation*.

22. Jordan Kahn (September 8, 2011). "Tim Cook Announces New Charity Matching Program for Apple Employees," *9 to 5 Mac*. http://9to5mac.com/2011/09/08/tim-cook-announces-new-charity-matching-serVice-for-apple-employees/

23. Leander Kahney (January 25, 2006). "Jobs vs. Gates: Who's the Star?" *Wired.* http://archive.wired.com/gadgets/mac/commentary/cultofmac/2006/01/70072

24. Tom Junod (October 2008). "Steve Jobs and the Portal to the Invisible," *Esquire.*

25. *In Business* (Nov/Dec 1995). "A Footnote That Was Worth $1 Billion."

26. BBC News (December 21, 2012). Steve Jobs' High-tech Yacht Impounded over Bill Dispute. http://www.bbc.com/news/business-20815910

27. Jemima Kiss (May 13, 2013). "Bill Gates: Steve Jobs and I Grew Up Together," *The Guardian.*

28. Walter Isaacson (2011). *Steve Jobs*. Simon & Schuster.

29. Leander Kahney (October 14, 2010). "John Sculley on Steve Jobs," *Cult of Mac.* http://www.cultofmac.com/63295/john-sculley-on-steve-jobs-the-full-interview-transcript/

30. Ibid.

31. David Hill (November 2011). "Steve Jobs: 'A Great Client,'" *Architectural Record.*

32. Sarah Amelar (April 2011). "Steve Jobs, the Demo Man," *Architectural Record.*

33. Federal Bureau of Investigation (n.d.). Steve Jobs's FBI File.

34. Felecia R. Lee (July 28, 1989). "121 Years of Men Only Ends at Club," *The New York Times.* http://www.nytimes.com/1989/07/28/nyregion/121-years-of-men-only-ends-at-club.html?src=pm

PART 6

1. Mona Simpson (October 30, 2011). "A Sister's Eulogy for Steve Jobs," *The New York Times*. http://www.nytimes.com/2011/10/30/opinion/mona-simpsons-eulogy-for-steve-jobs.html?_r=0

CHAPTER 14

1. David Sheff (February 1985). "Playboy Interview: Steven Jobs," *Playboy.* http://www.playboy.com/magazine/playboy-interview-steve-jobs

2. *WGBH* (May 14, 1990). Interview with Steve Jobs. http://openvault.wgbh.org/catalog/7b7ae3-steve-jobs-interview

3. Brent Schlender (November 9, 1998). "The Three Faces of Steve," *Fortune*. http://money.cnn.com/magazines/fortune/fortune_archive/1998/11/09/250880/index.htm

4. Steve Jobs (2005). Stanford Commencement Address.

5. Betsy Morris (February 2008). "Steve Jobs Speaks Out," *Fortune*. http://money.cnn.com/galleries/2008/fortune/0803/gallery.jobsqna.fortune/index.html

6. *NPR* (October 25, 2011). Jobs' Biography: Thoughts on Life, Death, and Apple.

7. Ibid.

8. Steve Lohr (October 20, 2011). "Jobs Tried Exotic Treatments to Combat Cancer, Book Says," *The New York Times*. http://www.nytimes.com/2011/10/21/technology/book-offers-new-details-of-jobs-cancer-fight.html

9. Dan Levine (October 7, 2011). "Who Will Inherit Steve Jobs' Estate?" *The Globe and Mail*.

10. Danielle Mayoras and Andy Mayoras (October 7, 2011). "Steve Jobs Appears to Have Protected His Estate with Living Trusts," *Forbes*. http://www.forbes.com/sites/trialandheirs/2011/10/07/steve-jobs-appears-to-have-protected-his-estate-with-living-trusts/

11. Peter Elkind (April 6, 2015). "When Steve Jobs's Ex-girlfriend Asked Him to Pay $25 Million for his 'Dishonorable behavior,'" *Fortune*. http://Fortune.com/2015/08/06/steve-jobs-apple-girlfriend/

12. Julia McKinnell (November 4, 2013). "Steve Jobs: Genius, and Lousy Father," *Maclean's*.

13. Ed Catmull and Amy Wallace (2014). *Creativity, Inc.: Overcoming the Unseen Forces That Stand in the Way of True Inspiration*. Random House.

14. Mary Riddell (January 27, 2012). "Bill Gates: 'I Wrote Steve Jobs a Letter as He Was Dying. He Kept It by His Bed,'" *The Telegraph*. http://www.telegraph.co.uk/technology/bill-gates/9041726/Bill-Gates-Iwrote-Steve-Jobs-a-letter-as-he-was-dying.-He-kept-it-by-his-bed.html.

15. Allie Townsend (October 11, 2011). "Chicago Ideas Week: What's in Bill Clinton's Head," *Time*. http://content.time.com/time/

specials/packages/article/0,28804,2096504_2096506_2096669,00.html

16. Mike Cassidy (October 10, 2011). "Cassidy: Woz and Jobs were Silicon Valley's Biggest Buddy Story," *San Jose Mercury News*. http://www.mercurynews.com/ci_19082667

17. Mona Simpson (October 30, 2011). "A Sister's Eulogy for Steve Jobs," *The New York Times*. http://www.nytimes.com/2011/10/30/opinion/mona-simpsons-eulogy-for-steve-jobs.html?_r=0

18. Bloomberg News (October 10, 2011). "Jobs' Cause of Death is Released by Coroner," *The New York Times*. http://www.nytimes.com/2011/10/11/business/steve-jobss-cause-of-death-is-released-by-coroner.html?_r=0

19. Shan Li (October 11, 2011). "Steve Jobs Buried in Alta Mesa among other Technologists," *The Los Angeles Times*. http://latimesblogs.latimes.com/technology/2011/10/steve-jobs-grave-alta-mesa.html

CHAPTER 15

1. United States Patent and Trademark Office (2009). Search for Patents. http://www.uspto.gov/patents-application-process/search-patents#heading-1

2. United States Patent and Trademark Office (2009). *Design Patent Application Guide*. http://www.uspto.gov/patents-getting-started/patent-basics/types-patent-applications/design-patent-application-guide

3. Chris Foresman (March 4, 2010). Apple vs HTC: Proxy Fight over Android Could Last Years. http://arstechnica.com/apple/2010/03/apple-vs-htc-a-proxy-fight-over-android-could-last-years/

4. *Apple* (April 15, 2011). Apple vs Samsung. https://www.apple.com/pr/pdf/110415samsungcomplaint.pdf

5. Brian X. Chen (November 30, 2014). "Star Witness in Apple Lawsuit Is Still Steve Jobs," *The New York Times*. http://www.nytimes.com/2014/12/01/technology/star-witness-in-apple-suit-is-steve-jobs.html?_r=0

6. Ibid.

7. Mark Ames (March 24, 2014). Newly Unsealed Documents Show Steve Jobs' Brutal Response after Getting a Google Employee Fired. https://pando.com/2014/03/25/newly-unsealed-documents-show-steve-jobs-brutally-callous-response-after-getting-a-google-employee-fired/

8. Ibid.

9. Ibid.

10. United States Department of Justice (September 24, 2010). *United States of America vs. Adobe, Apple, Google, Intel, Intuit, and Pixar*. https://www.justice.gov/atr/case-document/file/483451/download

11. Brian X. Chen (November 30, 2014). "Star Witness in Apple Lawsuit Is Still Steve Jobs," *The New York Times*. http://www.nytimes.com/2014/12/01/technology/star-witness-in-apple-suit-is-steve-jobs.html?_r=0

12. Ibid.

13. Brian X. Chen and Nicole Perlroth (November 21, 2014). "Settlement in Apple Case Over E-Books Is Approved," *The New York Times*. http://www.nytimes.com/2014/11/22/technology/judge-approves-450-million-settlement-in-apple-e-book-case.html

CHAPTER 16

1. Richard Stengel (October 17, 2011). "The Legacy of Steve Jobs," *Time*.

2. Eliana Dockterman (October 22, 2015). "Read the *TIME* Magazine Story That Plays a Key Role in *Steve Jobs*," *Time*. http://time.com/4063996/steve-jobs-movie-Time-magazine-profile/

3. Rick Tetzeli (March 18, 2015). "Tim Cook on Apple's Future: Everything Can Change Except Values," *Fast Company*. http://www.fastcompany.com/3042435/steves-legacy-tim-looks-ahead

4. Ian Parker (February 23, 2015). "The Shape of Things to Come," *New Yorker*.

5. Chris Preimesberger (October 17, 2011). "Millions Held Deep Respect for Steve Jobs and the Products He Helped Build for Apple" *eWeek*.

6. Chris Preimesberger (November 8, 2013). "eWeek at 30: Steve Jobs Returns in 1997 to Revive a Moribund Apple," *eWeek*.

7. Chris Preimesberger (October 17, 2011). "Millions Held Deep Respect for Steve Jobs and the Products He Helped Build for Apple" *eWeek*.

8. Walter Isaacson (2011). *Steve Jobs*. Simon & Schuster

9. Dara Kerr (August 12, 2013). *CNET*. Larry Ellison: Steve Jobs was 'Our Edison.' http://www.cnet.com/news/larry-ellison-steve-jobs-was-our-edison/#!

10. Steve Kovach (August 13, 2013). "Larry Ellison: Apple Won't Be Nearly as Successful Without Steve Jobs," *Business Insider*. http://finance.yahoo.com/news/larry-ellison-apple-wont-nearly-143153384.html

11. Dara Kerr (August 12, 2013). *CNET*. Larry Ellison: Steve Jobs was 'Our Edison.' http://www.cnet.com/news/larry-ellison-steve-jobs-was-our-edison/#!

12. Rhiannon Williams (August 30, 2015). "John Sculley: 'Steve Jobs was misrepresented in popular culture,'" *The Telegraph*. http://www.telegraph.co.uk/technology/2015/12/11/john-sculley-steve-jobs-was-misrepresented-in-popular-culture/

13. Marc Schneider (September 15, 2014). "Watch Tim Cook Talk Beats Music, Apple Watch & Steve Jobs' Untouched Office on *Charlie Rose*," *Billboard*. http://www.billboard.com/articles/business/6251519/tim-cook-charlie-rose-interview-apple-watch-beats

14. Alex Fitzpatrick (March 18, 2015). "Tim Cook Says This Was Steve Jobs' Biggest Gift to the World," *Time.com*.

15. Charlie Rose (December 20, 2015). *CBS News*. "What's Next for Apple."http://www.cbsnews.com/news/60-minutes-apple-tim-cook-charlie-rose/

16. Michael Simon (February 2016). "Cannibalism Is Good, and Other Things I Learned from 60 Minutes' Apple Report," *Macworld—Digital Edition*..

17. Charlie Rose (December 20, 2015). *CBS News*. What's Next for Apple http://www.cbsnews.com/news/60-minutes-apple-tim-cook-charlie-rose/

18. Mike Godwin (February 2012). "Steve Jobs, the Inhumane Humanist," *Reason*.

19. Lev Grossman (March 25, 2015). "Why It Matters Who Steve Jobs Really Was," *Time.com*.

20. Don Steinberg (September 30, 2015). "'Steve Jobs' Seen Through Product Launches," *The Wall Street Journal.* http://www.wsj.com/articles/steve-jobs-seen-through-product-launches-1443626394

21. Shira Ovide (September 18, 2014). "Larry Ellison to Step Aside as Oracle CEO," *The Wall Street Journal.* http://www.wsj.com/articles/larry-ellison-to-step-aside-as-oracle-ceo-1411070636

22. Li Yuan (October 7, 2015). "A Steve Jobs Disciple Dreams Big," *The Wall Street Journal.* http://www.wsj.com/articles/a-steve-jobs-disciple-dreams-big-1444246380

23. Rana Foroohar (February 27, 2012). "What Would Steve Do?" *Time.*

24. Ibid.

APPENDIX

1. The United States Patent and Trademark Office (2014). The National Medal of Technology and Innovation 1985 Laureates. http://www.uspto.gov/about/nmti/recipients/1985.jsp.

2. *Jefferson Awards Foundation* (n.d.) Past Winners. http://www.jeffersonawards.org/past-winners/

3. Bo Burlingham and George Gendron (April 1, 1989). "The Entrepreneur of the Decade," *Inc Magazine.*

4. California Museum (2016). *Steve Jobs.* http://www.californiamuseum.org/inductee/steve-jobs

5. California Museum (2016). California Museum FAQ. http://www.californiamuseum.org/faqs-0

6. *The Recording Academy* (2016). Trustees Award. https://www.grammy.org/recording-academy/awards/trustee-awards

7. Ricky Brigante (July 10, 2013). "Steve Jobs, Dick Clark, Billy Crystal, John Goodman among Disney Legends Awards Recipients Announced for 2013," *Inside the Magic.* D23 Expo. http://www.insidethemagic.net/2013/07/steve-jobs-dick-clark-billy-crystal-john-goodman-among-disney-legends-awards-recipients-announced-for-2013-d23-expo/

INDEX

About the Author

Michael B. Becraft, DMgt, is the former Edward F. Lyle Professor of Finance and director of the Graduate Program in Business at Park University in Kansas City, MO. He is the author of Greenwood's *Bill Gates: A Biography* and contributed to Greenwood's *Economic Thinkers: A Biographical Encyclopedia.*